# Simplifying Data Engineering and Analytics with Delta

Create analytics-ready data that fuels artificial
intelligence and business intelligence

**Anindita Mahapatra**

BIRMINGHAM—MUMBAI

# Simplifying Data Engineering and Analytics with Delta

Copyright © 2022 Packt Publishing

**Publishing Product Manager**: Dhruv Jagdish Kataria
**Senior Editor**: Tazeen Shaikh
**Content Development Editor**: Sean Lobo, Priyanka Soam
**Technical Editor**: Devanshi Ayare
**Copy Editor**: Safis Editing
**Project Coordinator**: Farheen Fathima
**Proofreader**: Safis Editing
**Indexer**: Manju Arasan
**Production Designer**: Roshan Kawale
**Marketing Coordinator**: Nivedita Singh

First published: July 2022

Production reference: 2290722

Published by Packt Publishing Ltd.

Livery Place
35 Livery Street
Birmingham
B3 2PB, UK.

ISBN 978-1-80181-486-7

www.packt.com

*This book is dedicated to my parents for their unconditional love and support.*

*While there are too many to name here, I would like to thank my mentors and colleagues that have encouraged and aided me in this journey. Last but not least, I would like to thank the team at Packt for all their help and guidance throughout the process.*

# Foreword

My father was one of the first **chief information officers** (**CIOs**) back in the mid-1980s. He led all of IT for the largest commercial property insurer in the world. He reported to the CEO, which, at that time, was uncommon as most IT functions reported to the CFO because they were cost centers. Every weekend he would bring home some type of new technology: an Apple 2E, an IBM PC, even a "portable" computer that weighed 40 lbs. My sisters and I would play with them for hours on end, creating spreadsheets and writing basic programs. At the time, I viewed him as being on the bleeding edge of technology, a real "techie."

When I graduated college and went to work at IBM in 1991, I came home and tried to talk about technology with my father using all the speeds and feeds of the mid-range and Unix systems that I had just been trained on. Each time I mentioned a particular technical specification, he would ask me "*What does that do*?" or "*Why is that important?*" His questions frustrated me. When I explained why the SPEC-INT metric was important, he would look confused. I began to think my father wasn't the techie I once believed him to be. And I was right. Part of me was disappointed with this realization. But, over time, I came to see that his expertise was not the technology itself, but understanding the business strategy deeply and translating how specific capabilities provided by technology could be applied to make the business strategy succeed.

Fast forward 30+ years and I'm now the vice president of Global Value Acceleration at Databricks, one of the fastest-growing software companies in history. I lead a global team of consultants, or translators, that help prospects and customers connect the technical power of our data and AI platform to the meaningful business value its capabilities will deliver as they pursue their business strategy.

Looking back, I realize that I've been doing value translation my entire career. I found that when the business strategy meets the technical strategy and they are well aligned, magic happens. Executives who hold budgets and decision-making authority accelerate and approve initiatives and their associated spending. Likewise, when the translation work isn't done or isn't done well, they deny those requests. Over my career, I've learned that when those requests fail, it's generally not the fault of the technology. It comes down to the quality of the translation and the underlying story.

The need for translators in data is significant and increasing. According to a recent McKinsey article, "(data) translators play a critical role in bridging the technical expertise of data engineers and data scientists with the operational expertise of marketing, supply chain, manufacturing, risk, and other frontline managers. In their role, translators help ensure that the deep insights generated through sophisticated analytics translate into impact at scale in an organization. By 2026, the McKinsey Global Institute estimates that demand for translators in the United States alone may reach two to four million."

Through thousands of translation engagements with global enterprises over the last decade, my team, with our **business value assessment** (**BVA**) methodology, has proven to be a critical ingredient to the success of large initiatives. The recipe that translates complex technology to the C-suite for investment consideration follows a simple framework comprising a story. It draws executives in, making it easy for them to say "yes":

1. Key strategic priorities
2. Use cases aligned with those priorities
3. Technical barriers in the way of success
4. Capabilities required to succeed
5. Value to be realized when successful
6. Return on investment
7. Success plan

According to **International Data Corporation** (**IDC**), 95% of technology investments require financial justification. This framework provides the financial justification that is needed, but it also reinforces the urgency to act by connecting the project to the most important priorities or business problems that the C-suite and board have their eyes on, and it specifies the capabilities required for success. When you put these together, you have a CFO-ready business case that qualifies and quantifies the value setting your project apart from all others.

This is why I've been so excited about this book. The opportunity to apply powerful technology such as Delta and deliver impact all the way to the boardroom of your employer is real and required for success in today's market. When I first worked with Anindita at Databricks, it was clear to me that she has a special talent that few technical people have. She is a translator. She can speak succinctly about very complex technical topics, make them easy to understand at any level, and connect the technology to why it matters to the business. Her ability to do this for our customers and for other Databricks employees has helped her, and Databricks, succeed in many ways.

As you read on from here, note how everything from data modeling to operationalizing Delta pipelines is made easy to understand and translatable to the business. Anindita, in her special way, will guide you to become a better data engineer while infusing you with specific skills to become a data translator, whose future value may just be priceless.

*Doug May*

*VP, Global Value Acceleration*

*Databricks Inc.*

# Contributors

## About the author

**Anindita Mahapatra** is a lead solutions architect at Databricks in the data and AI space helping clients across all industry verticals reap value from their data infrastructure investments. She teaches a data engineering and analytics course at Harvard University as part of their extension school program. She has extensive big data and Hadoop consulting experience from Think Big/Teradata, prior to which she was managing the development of algorithmic app discovery and promotion for both Nokia and Microsoft stores. She holds a master's degree in liberal arts and management from Harvard Extension School, a master's in computer science from Boston University, and a bachelor's in computer science from BITS Pilani, India.

# About the reviewer

**Oleksandra Bovkun** is a solutions architect for data and AI platforms and systems. She works with customers and engineering teams to develop architectures and solutions for data platforms and guide them through the implementation. She has extensive experience and expertise in open source technologies such as Apache Spark, MLflow, Delta Lake, Kubernetes, and Helm, and programming languages such as Python and Scala. Furthermore, she specializes in data platform architecture, especially in DevOps and MLOps processes. Oleksandra has more than 10 years of experience in the field of software development and data engineering. She likes to discover new technologies and tools, architecture patterns, and open source projects.

# Table of Contents

# 3

# Delta – The Foundation Block for Big Data

# Section 2 – End-to-End Process of Building Delta Pipelines

# 4

# Unifying Batch and Streaming with Delta

# 5

# Data Consolidation in Delta Lake

# 6

# Solving Common Data Pattern Scenarios with Delta

# 7
# Delta for Data Warehouse Use Cases

# 8
# Handling Atypical Data Scenarios with Delta

# 9
# Delta for Reproducible Machine Learning Pipelines

# 10
# Delta for Data Products and Services

# Section 3 – Operationalizing and Productionalizing Delta Pipelines

# 11
# Operationalizing Data and ML Pipelines

# 12

# Optimizing Cost and Performance with Delta

# 13

# Managing Your Data Journey

## Index

## Other Books You May Enjoy

# Preface

Delta helps you generate reliable insights at scale and simplifies architecture around data pipelines, allowing you to focus primarily on refining the use cases being worked upon. This is especially important considering the same architecture is reused when onboarding new use cases.

In this book, you'll learn the principles of distributed computing, data modeling techniques, big data design patterns, and templates that help solve end-to-end data flow problems for common scenarios and are reusable across use cases and industry verticals. You'll also learn how to recover from errors and the best practices around handling structured, semi-structured, and unstructured data using Delta. Next, you'll get to grips with features such as ACID transactions on big data, disciplined schema evolution, time travel to help rewind a dataset to a different time or version, and unified batch and streaming capabilities that will help you build agile and robust data products.
By the end of this book, you'll be able to use Delta as the foundational block for creating analytics-ready data that fuels all AI/BI use cases.

> **Delta 2.0**
>
> Delta Lake 2.0, the latest release of Delta Lake announced at DAIS 2022 will further enable everyone to benefit from all Delta Lake innovations with all Delta Lake APIs being open-sourced — in particular, the performance optimizations and functionality brought on by Delta Engine like ZOrder (https://docs.delta.io/2.0.0/optimizations-oss.html#z-ordering-multi-dimensional-clustering), Change Data Feed (`https://docs.delta.io/2.0.0/delta-change-data-feed.html`), Dynamic Partition Overwrites (https://docs.delta.io/2.0.0/delta-batch.html#dynamic-overwrites), and Dropped Columns.

# Who this book is for

Individuals in the data domain such as data engineers, data scientists, ML practitioners, and BI analysts working with big data will be able to put their knowledge to work with this practical guide to executing pipelines and supporting diverse use cases using the Delta protocol. Basic knowledge of SQL, Python programming, and Spark is required to get the most out of this book.

# What this book covers

*Chapter 1*, *Introduction to Data Engineering*, covers how data is the new oil. Just as oil has to burn to get heat and light, data also has to be harnessed to get valuable insights. The quality of insights will depend on the quality of the data. So, understanding how to manage data is an important function for every industry vertical. This chapter introduces the fundamental principles of data engineering and addresses the growing trends in the industry of data-driven organizations and how to leverage IT data operation units as a competitive advantage instead of viewing them as a cost center.

*Chapter 2*, *Data Modeling and ETL*, covers how leveraging the scalability and elasticity of the cloud helps turn on compute on demand and move CAPEX allocation towards OPEX. This chapter introduces common big data design patterns and best practices for modeling big data.

*Chapter 3*, *Delta – The Foundational Block for Big Data*, introduces Delta as a file format and points out features that Delta brings to the table over vanilla Parquet and why it is a natural choice for any pipeline. Delta is an overloaded term – it is a protocol first, a table next, and a lake finally!

*Chapter 4*, *Unifying Batch and Streaming with Delta*, covers how the trend is towards real-time ingestion, analysis, and consumption of data. Batching is actually a type of streaming workload. Reader/writer isolation is necessary in an environment with multiple producers/consumers involving the same data assets to work independently with the promise that bad or partial data is never presented to the user.

*Chapter 5*, *Data Consolidation in Delta Lake*, covers how bringing data together from various silos is only the first step towards building a data lake. The real deal is in increased reliability, quality, and governance, which needs to be enforced to get the most out of the data and infrastructure investment while adding value to any BI or AI use case built on top of it.

*Chapter 6*, *Solving Common Data Pattern Scenarios with Delta*, covers common CRUD operations on big data and looks at use cases where they can be applied as a repeatable blueprint.

*Chapter 7, Delta for Data Warehouse Use Cases*, covers the journey from databases to data warehouses to data lakes, and finally, to lakehouses. The unification of data platforms has never been more important. Is it possible to house all kinds of use cases with a single architecture paradigm? This chapter focuses on the data handling needs and capability requirements that drive the next round of innovation.

*Chapter 8, Handling Atypical Data Scenarios with Delta*, covers several conditions, such as data imbalance, skew, and bias, that need to be addressed to ensure data is not only cleansed and transformed per the business requirements but is also conducive to the underlying compute and for the use case at hand. Even when the logic of the pipelines has been ironed out, there are other statistical attributes of the data that need to be addressed to ensure that the data characteristics for which it was initially designed still hold and make the most of the distributed compute.

*Chapter 9, Delta for Reproducible Machine Learning Pipelines*, emphasizes that if ML is hard, then reproducible ML and productionizing of ML is even harder. A large part of ML is data preparation. The quality of insights will be as good as the quality of the data that is used to build the models. In this chapter, we look at the role of Delta in ensuring reproducible ML.

*Chapter 10, Delta for Data Products and Services*, covers consumption patterns of data democratization that ensure the curated data gets into the hands of the consumers in a timely and secure manner so that the insights can be leveraged meaningfully. Data can be served both as a product and a service, especially in the context of a mesh architecture involving multiple lines of businesses specializing in different domains.

*Chapter 11, Operationalizing Data and ML Pipelines*, looks at the aspects of a mature pipeline that make it considered production worthy. A lot of the data around us remains in unstructured form and carries a wealth of information, and integrating it with more structured transactional data is where firms can not only get competitive intelligence but also begin to get a holistic view of their customers to employ predictive analytics.

*Chapter 12, Optimizing Cost and Performance with Delta*, looks at how running a pipeline faster has cost implications that translate directly to increased infrastructural savings. This applies to both the ETL pipeline that brings in the data and curates it as well as the consumption pipeline where the stakeholders tap into this curated data. In this chapter, we look at strategies such as file skipping, z-ordering, small file coalescing, and bloom filtering to improve query runtime.

*Chapter 13*, *Managing Your Data Journey*, emphasizes the need for policies around data access and data use that need to be honored as per regulatory and compliance guidelines. In some industries, it may be necessary to provide evidence of all data access and transformations. Hence, there is a need to be able to set controls in place, detect if something has been changed, and provide a transparent audit trail.

# To get the most out of this book

Basic knowledge of SQL, Python programming, and Spark is required to get the most out of this book. Delta is open source and can be run both on-prem and in the cloud. Because of the rise in cloud data platforms, a lot of the descriptions and examples are in the context of cloud storage.

Use the following GitHub link for the Delta Lake documentation and quickstart guide to help you set up your environment and become familiar with the necessary APIs: `https://github.com/delta-io/delta`.

Databricks is the original creator of Delta, which was open sourced to the Linux Foundation and is supported by a large user community. Examples in this book cover some Databricks-specific features to provide a complete view of features and capabilities. Newer features continue to be ported from Databricks to open source Delta. Please refer to the proposed roadmap for the feature migration details: `https://github.com/delta-io/delta/issues/920`.

**If you are using the digital version of this book, we advise you to type the code yourself or access the code from the book's GitHub repository (a link is available in the next section). Doing so will help you avoid any potential errors related to the copying and pasting of code.**

# Download the example code files

You can download the example code files for this book from GitHub at `https://github.com/PacktPublishing/Simplifying-Data-Engineering-and-Analytics-with-Delta`.

If there's an update to the code, it will be updated in the GitHub repository.

We also have other code bundles from our rich catalog of books and videos available at `https://github.com/PacktPublishing/`. Check them out!

# Download the color images

We also provide a PDF file that has color images of the screenshots and diagrams used in this book. You can download it here: `https://packt.link/UI11F`.

# Conventions used

There are a number of text conventions used throughout this book.

`Code in text`: Indicates code words in text, database table names, folder names, filenames, file extensions, pathnames, dummy URLs, user input, and Twitter handles. Here is an example: "There is no need to run the `REPAIR TABLE` command when you're working with the Delta format".

A block of code is set as follows:

```
SELECT COUNT(*) FROM some _ parquet _ table
```

**Bold**: Indicates a new term, an important word, or words that you see onscreen. For instance, words in menus or dialog boxes appear in **bold**. On the other hand, a **data swamp** is a large body of data that is ungoverned and unreliable.

> **Tips or Important Notes**
> Appear like this.

# Get in touch

Feedback from our readers is always welcome.

**General feedback**: If you have questions about any aspect of this book, email us at `customercare@packtpub.com` and mention the book title in the subject of your message.

**Errata**: Although we have taken every care to ensure the accuracy of our content, mistakes do happen. If you have found a mistake in this book, we would be grateful if you would report this to us. Please visit `www.packtpub.com/support/errata` and fill in the form.

**Piracy**: If you come across any illegal copies of our works in any form on the internet, we would be grateful if you would provide us with the location address or website name. Please contact us at copyright@packt.com with a link to the material.

**If you are interested in becoming an author**: If there is a topic that you have expertise in and you are interested in either writing or contributing to a book, please visit authors.packtpub.com.

## Share Your Thoughts

Once you've read *Simplifying Data Engineering and Analytics with Delta*, we'd love to hear your thoughts! Scan the QR code below to go straight to the Amazon review page for this book and share your feedback.

https://packt.link/r/1-801-81486-4

Your review is important to us and the tech community and will help us make sure we're delivering excellent quality content.

# Section 1 – Introduction to Delta Lake and Data Engineering Principles

Understanding modern data architectures and sound data engineering principles and practices are crucial to ensure that your AI and BI strategies are reliable and defensible. Generated insights are going to be as good as the quality of the underlying data, so the upfront effort put into understanding the data, modeling it, and transforming it per the business needs goes a long way to foster innovation, productivity, and agility in your data teams.

This part includes the following chapters:

- *Chapter 1, An Introduction to Data Engineering*
- *Chapter 2, Data Modeling and ETL*
- *Chapter 3, Delta – The Foundation Block for Big Data*

# 1
# Introduction to Data Engineering

*"Water, water, everywhere, nor any drop to drink...*

*Data data everywhere, not a drop of insight!"*

With the vast exodus of data around us, it is important to crunch it meaningfully and promptly to extract value from all the noise. This is where data engineering steps in. If collecting data is the first step, drawing useful insights is the next. **Data engineering** encompasses several personas that come together with their unique individual skill sets and processes to bring this to fruition. Data usually outlives the technology, and it continues to grow. New tools and frameworks come to the forefront to solve a lot of old problems. It is important to understand business requirements, the accompanying tech challenges, and typical shifts in paradigms to solve these age-old problems in a better manner.

By the end of this chapter, you should have an appreciation of the data landscape, the players, and the advances in distributed computing and cloud infrastructure that make it possible to support the high pace of innovation.

In this chapter, we will cover the following topics:

- The motivation behind data engineering
- Data personas
- Big data ecosystem
- Evolution of data stores
- Trends in distributed computing
- Business justification for tech spending

# The motivation behind data engineering

Data engineering is the process of converting *raw data* into *analytics-ready data* that is more accessible, usable, and consumable than its raw format. Modern companies are increasingly becoming *data-driven*, which means they use data to make business decisions to give them better *insights* into their customers and business operations. They can use these to improve profitability, reduce costs, and give them a competitive edge in the market. Behind the scenes, a series of tasks and processes are performed by a host of data personas who build reliable pipelines to source, transform, and analyze data so that it is a repeatable and mostly automated process.

Different systems produce different datasets that need to function as individual units and are brought together to provide a holistic view of the state of the business – for example, a customer buying merchandise through different channels such as the web, in-app, or in-store. Analyzing activity in all the channels will help predict the next customer purchase and possibly the next channel type as well. In other words, having all the datasets in one place can help answer questions that couldn't be answered by the individual systems. So, data consolidation is an industry trend that breaks down individual silos. However, each of the systems may have been designed differently, as well as different requirements and **service-level agreements** (**SLAs**), and now all of that needs to be normalized and consolidated in a single place to facilitate better analytics.

The following diagram compares the process of farming to that of processing and refining data. In both setups, there are different producers and consumers and a series of refining and packaging steps:

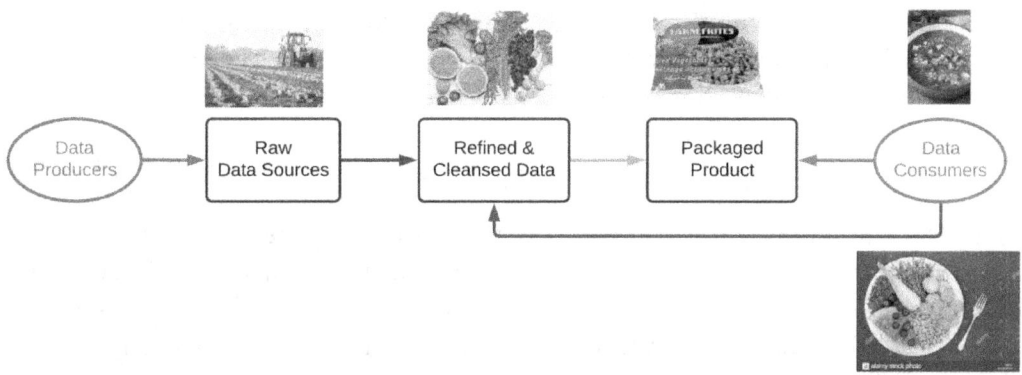

Figure 1.1 – Farming compared to a data pipeline

In this analogy, there is a farmer, and the process consists of growing crops, harvesting them, and making them available in a grocery store. This produce eventually becomes a ready-to-eat meal. Similarly, a data engineer is responsible for creating *ready-to-consume* data so that each consumer does not have to invest in the same heavy lifting. Each cook taps into different points of the pipeline and makes different recipes based on the specific needs of the use cases that need to be catered for. However, the freshness and quality of the produce are what make for a delightful meal, irrespective of the recipe that's used.

We are at the interesting conjunction of big data, the cloud, and **artificial intelligence** (**AI**), all of which are fueling tremendous innovation in every conceivable industry vertical and generating data exponentially. Data engineering is increasingly important as data drives business use cases in every industry vertical. You may argue that data scientists and machine learning practitioners are the unicorns of the industry, and they can work their magic for business. That is certainly a stretch of the imagination. Simple algorithms and a lot of good reliable data produce better insights than complicated algorithms with inadequate data. Some examples of how pivotal data is to the very existence of some of these businesses are listed in the following section.

## Use cases

In this section, we've taken a popular use case from a few industry verticals to highlight how data is being used as a driving force for their everyday operations and the scale of data involved:

- **Security Incident and Event Management** (**SIEM**) cyber security systems for threat detection and prevention.

This involves user activity monitoring and auditing for suspicious activity patterns and entails collecting a large volume of logs across several devices and systems, analyzing them in real time, correlating data, and reporting on findings via alerts and dashboard refreshes.

- Genomics and drug development in health and life sciences.

  The Human Genome project took almost 15 years to complete. A single human genome requires about 100 gigabytes of storage, and it is estimated that by 2025, 40 exabytes of data will be required to process and store all the sequenced genomes. This data helps researchers understand and develop cures that are more targeted and precise.

- Autonomous vehicles.

  Autonomous vehicles use a lot of unstructured image data that's been generated from cameras on the body of the car to make safe driving decisions. It is estimated that an active vehicle generates about 5 TB every hour. Some of it will be thrown away after a decision has been made, but a part of it will be saved both locally as well as transmitted to a data center for long-term trend monitoring.

- IoT sensors in Industry 4.0 smart factories in manufacturing.

  Smart manufacturing and the Industry 4.0 revolution, which are powered by advances in IoT, are enabling a lot of efficiencies in machine and human utilization on the shop floor. Data is at the forefront of scaling these smart factory initiatives with real-time monitoring, predictive maintenance, early alerting, and digital twin technology to create closed-loop operations.

- Personalized recommendations in retail.

  In an omnichannel experience, personalization helps retailers engage better with their customers, irrespective of the channel they choose to engage with, all while picking up the relevant state from the previous channel they may have used. They can address concerns before the customer churns to a competitor. Personalization at scale can not only deliver a percentage lift in sales but can also reduce marketing and sales costs.

- Gaming/entertainment.

  Games such as Fortnite and Minecraft have captivated children and adults alike who spend several hours in a multi-player online game session. It is estimated that Fortnite generates 100 MB of data per user, per hour. Music and video streaming also rely a lot on recommendations for new playlists. Netflix receives more than a million new ratings every day and uses several parameters to bin users to understand similarities in their preferences.

- Smart agriculture.

  The agriculture market in North America is estimated to be worth 6.2 billion US dollars and uses big data to understand weather patterns for smart irrigation and crop planting, as well as to check soil conditions for the right fertilizer dose. John Deere uses computer vision to detect weeds and can localize the use of sprays to help preserve the quality of both the environment and the produce.

- Fraud detection in the Fintech sector.

  Detecting and preventing fraud is a constant effort as fraudsters find new ways to game the system. Because we are constantly transacting online, a lot of digital footprints are left behind. By some estimates, about 10% of insurance company payments are made due to fraud. AI techniques such as biometric data and ML algorithms can detect unusual patterns, which leads to better monitoring and risk assessment so that the user can be alerted before a lot of damage is done.

- Forecasting use cases across a wide variety of verticals.

  Every business has some need for forecasting, either to predict sales, stock inventory, or supply chain logistics. It is not as straightforward as projection – other patterns influence this, such as seasonality, weather, and shifts in micro or macro-economic conditions. Data that's been augmented over several years by additional data feeds helps create more realistic and accurate forecasts.

## How big is big data?

90% of the data that's generated thus far has been generated in the last 2 years alone. At the time of writing, it is estimated that *2.5 quintillion (18 zeros)* bytes of data is produced every day. A typical commercial aircraft generates 20 terabytes of data per engine every hour it's in flight.

We are just at the beginning stages of autonomous driving vehicles, which rely on data points to operate. The world's population is about 7.7 billion. The number of connected devices is about 10 billion, with portions of the world not yet connected by the internet. So, this number will only grow as the exodus of IoT sensors and other connected devices grows. People have an appetite to use apps and services that generate data, including search functionalities, social media, communication, services such as YouTube and Uber, photo and video services such as Snapchat and Facebook, and more. The following statistics give you a better idea of the data that's generated all around us and how we need to swim effectively through all the waves and turbulences that they create to digest the most useful nuggets of information.

Every minute, the following occurs (approximately):

- 16 million text messages
- 1 million Tinder swipes
- 160 million emails
- 4 million YouTube videos
- 0.5 million tweets
- 0.5 million Snapchat shares

With so much data being generated, there is a need for robust data engineering tools and frameworks and reliable data and analytics platforms to harness this data and make sense of it. This is where data engineering comes to the rescue. Data is as important an asset as code is, so there should be governance around it. Structured data only accounts for 5-10% of enterprise data; semi-structured and unstructured data needs to be added to complete this picture.

Data is the new oil and is at the heart of every business. However, raw data by itself is not going to make a dent in a business. It is the useful insights that are generated from curated data that are the refined consumable oil that businesses aspire for. Data drives ML, which, in turn, gives businesses their competitive advantage. This is the age of *digitization*, where most successful businesses see themselves as tech companies first. Start-ups have the advantage of selecting the latest digital platforms while traditional companies are all undergoing digital transformations. Why should I care so much for the underlying data? I have highly qualified ML practitioners who are the unicorns of the industry that can use sophisticated algorithms and their special skill sets to make magic!

In this section, we established the importance of curating data since raw data by itself isn't going to make a dent in a business. In the next section, we will explore the influence that curated data has on the effectiveness of ML initiatives.

## But isn't ML and AI all the rage today?

AI and ML are catchy buzzwords, and everybody wants to be on the bandwagon and use ML to differentiate their product. However, the hardest part about ML is not ML – it is managing everything else around ML creation. This is shown by Google in one of their papers in 2014 (`https://papers.nips.cc/paper/2015/file/86df7dcfd89` `6fcaf2674f757a2463eba-Paper.pdf`). Garbage in, garbage out, is true. The magic wand of ML will only work if the boxes surrounding it are well developed and most of them are data engineering tasks. In short, high-quality curated data is the foundational layer of any ML application, and the data engineering practices that curate this data are the backbone that holds it all together:

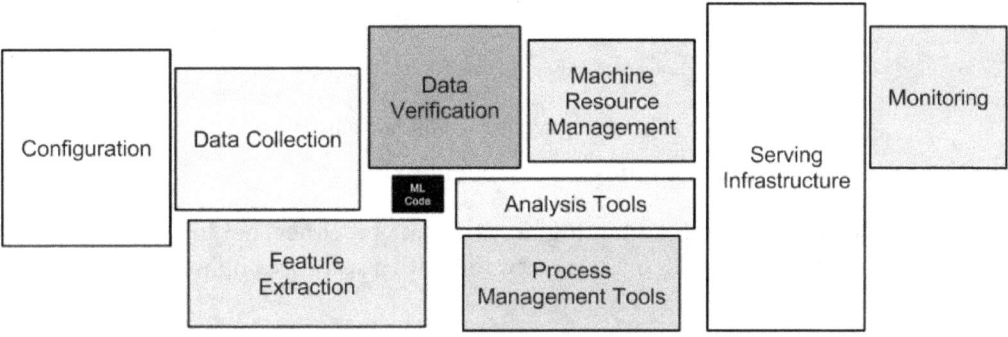

Figure 1.2 – The hardest part about ML is not ML, but rather everything else around it

Technologies come and go, so understanding the core challenges around data is critical. As technologists, we create more impact when we align solutions with business challenges. Speed to insights is what all businesses demand and the key to this is data. The data and IT functional areas within an organization that were traditionally viewed as cost centers are now being viewed as revenue-generating sources. Organizations where business and tech cooperate, instead of competing with each other, are the ones most likely to succeed with their data initiatives. Building data services and products involves several personas. In the next section, we will articulate the varying skill sets of these personas within an organization.

# Understanding the role of data personas

Since data engineering is such a crucial field, you may be wondering who the main players are and what skill sets they possess. Building a data product involves several folks, all of whom need to come together with seamless handoffs to ensure a successful end product or service is created. It would be a mistake to create silos and increase both the number and complexity of integration points as each additional integration is a potential failure point. Data engineering has a fair overlap with software engineering and data science tasks:

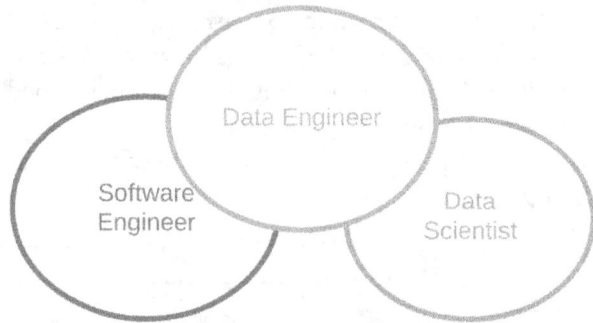

Figure 1.3 – Data engineering requires multidisciplinary skill sets

All these roles require an understanding of data engineering:

- **Data engineers** focus on maintaining how the data pipelines that ingest and transform data run. This has a lot in common with a software engineering role coupled with lots of data.

- **BI analysts** focus on SQL-based reporting and can be operational or domain-specific **subject-matter experts** (**SMEs**) such as financial or supply chain analysts.

- **Data scientists and ML practitioners** are statisticians who explore and analyze the data (via **Exploratory Data Analysis** (**EDA**)) and use modeling techniques at various levels of sophistication.

- **DevOps and MLOps** focus on the infrastructure aspects of monitoring and automation. MLOps is DevOps coupled with the additional task of managing the life cycle of analytic models.

- **ML engineers** refer to folks who can span across both the data engineer and data scientist roles.

- **Data leaders** are chief data officers – that is, data stewards who are at the top of the food chain in terms of the ultimate governors of data.

The following diagram shows the typical placement of the four main data personas working collaboratively on a data platform to produce business insights to give the company a competitive advantage in the industry:

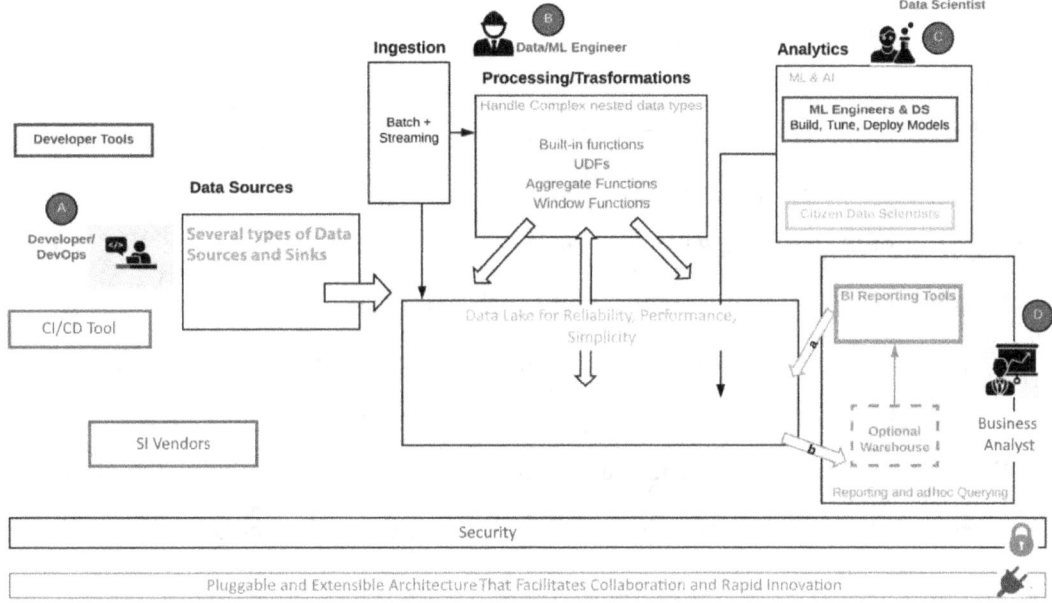

Figure 1.4 – Data personas working in collaboration

Let's take a look at a few of these points in more detail:

A. DevOps is responsible for ensuring all operational aspects of the data platform and traditionally does a lot of scripting and automation.

B. Data/ML engineers are responsible for building the data pipeline and taking care of the **extract, transform, load** (**ETL**) aspects of the pipeline.

C. Data scientists of varying skill levels build models.

D. Business analysts create reporting dashboards from aggregated curated data.

# Big data ecosystem

The big data ecosystem has a fairly large footprint that's contributed by several infrastructures, analytics (BI and AI) technologies, data stores, and apps. Some of these are open source, while others are proprietary. Some are easy to wield, while others have steeper learning curves. Big data management can be daunting as it brings in another layer of challenges over existing data systems. So, it is important to understand what qualifies as a big data system and know what set of tools should be used for the use case at hand.

# What characterizes big data?

Big data was initially characterized with three Vs (volume, velocity, and variety). This involves processing a lot of data coming into a system at high velocity with varying data types. Two more Vs were subsequently added (veracity and value). This list continues to grow and now includes variability and visibility. Let's look at the top five and see what each of them mean:

- **Volume**: This is measured by the size of data, both historical and current:

  - The number of records in a file or table

  - The size of the data in gigabytes, terabytes, and so on

- **Velocity**: This refers to the frequency at which new data arrives:

  - The batches have a well-defined interval, such as daily or hourly.

  - Real time is either continuous or micro-batch, typically in seconds.

- **Variety**: This refers to the structural nature of the data:

  - Structured data is usually relational and has a well-defined schema.

  - Semi-structured data has a self-describing schema that can evolve, such as the XML and JSON formats.

  - Unstructured data refers to free-text documents, audio, and video data that's usually in binary format.

- **Veracity**: This refers to the trustworthiness and reliability of the data:

  - Lineage refers to not just the source but also the subsequent systems where transformations took place to ensure that data fidelity is maintained and can be audited. To guarantee such reliability, data lineage must be maintained.

- **Value**: This refers to the business impact that the dataset has – that is, how valuable the data is to the business.

# Classifying data

Different classification gauges can be used. The common ones are based on the following aspects:

- As the volume of data increases, we move from regular systems to big data systems. Big data is typically terabytes of data that cannot fit on a single computer node.

- As the velocity of the data increases, we move toward big data systems specialized in streaming. In batch systems, irrespective of when data arrives, it is processed at a predefined regular interval. In streaming systems, there are two flavors. If it's set to continuous, data is processed as it comes. If it's set to micro-batch, data is aggregated in small batches, typically a few seconds or milliseconds.

- When it comes to variety – that is, the structure of the data – we move toward the realm of big data systems. In structured data, the schema is well known and stable, so it's assumed to be fairly static and rigid to the definition. With semi-structured data, the schema is built into the data and can evolve. In unstructured data such as images, audio, and video, there is some metadata but no real schema to the binary data that's sent.

The following diagram shows what trends in data characteristics signal a move toward big data systems. For example, demographic data is fairly structured with predefined fields, operational data moves toward the semi-structured realm as schemas evolve, and the most voluminous is behavioral data as it encompasses user sentiment, which is constantly changing and is best captured by unstructured data such as text, audio, and images:

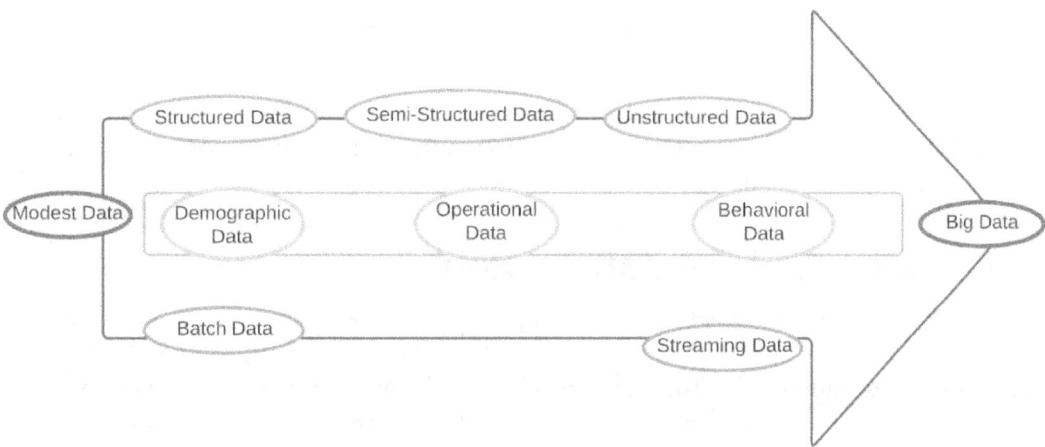

Figure 1.5 – Classifying data

Now that we have covered the different types of data, let's see how much processing needs to be done before it can be consumed.

# Reaping value from data

As data is refined and moves further along the pipeline, there is a tradeoff between the value that's added and the cost of the data. In other words, more time, effort, and resources are used, which is why the cost increases, but the value of the data increases as well:

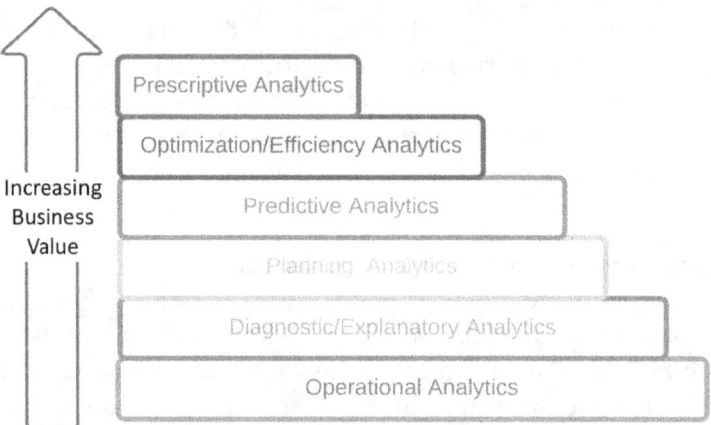

Figure 1.6 – The layers of data value

The analogy we're using here is that of cutting carbon to create a diamond. The raw data is the carbon, which gets increasingly refined. The longer the processing layers, the more refined and curated the value of the data. However, it is more time-consuming and expensive to produce the artifact.

# Top challenges of big data systems

People, technology, and processes are the three prongs that every enterprise has to keep up with. Technology changes around us at a pace that is hard to keep up with and gives us better tools and frameworks. Tools are great but until you train people to use them effectively, you cannot create solutions, which is what a business needs. Sound and effective business processes help you pass information quickly and break data silos.

According to Gartner, the three main challenges of big data systems are as follows:

- Data silos
- Fragmented tools
- People with the skill sets to wield them

The following diagram shows these challenges:

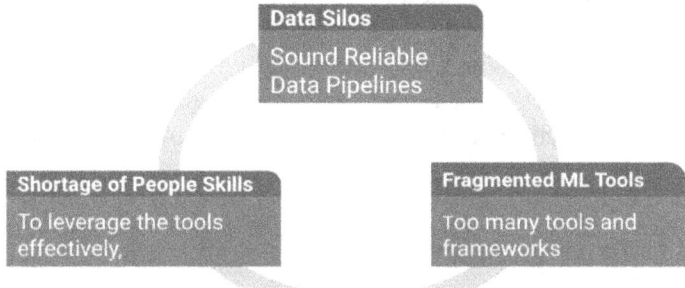

Figure 1.7 – Big data challenges

Any imbalance or immaturity in these areas results in poor insights. These challenges around data quality and data staleness lead to inaccurate, delayed, and hence unusable insights.

# Evolution of data systems

We have been collecting data for decades. The flat file storages of the 60s led to the data warehouses of the 80s to **Massively Parallel Processing** (**MPP**) and NoSQL databases, and eventually to data lakes. New paradigms continue to be coined but it would be fair to say that most enterprise organizations have settled on some variation of a data lake:

| 1960s | 1980s | 2000s | 2010s | 2020s |
|---|---|---|---|---|
| **Start of DBMS Technologies** | **Data Warehouses** | **Web & Unstructured Data** | **Data Lakes** | **Lakehouse** |
| | The 1990s saw the rise of Data Warehouses, Dimensional Modeling, Data Marts | Audio, Video Codecs exploded. Emphasis on Metadata grew. Streaming requirements surfaced | Spark increased in popularity and adoption because of speed and agility. | Data Mesh, Data Fabric, Lakehouse are the newer entrants |
| Staring with the flat files in the 60s and moving on to DBMS in 70s | This also saw the rise of MPP databases (such as Teradata) | NoSQL databases came to handle processing needs | Move to Cloud Data Platforms with cheaper storage. | Focus on Data Domains & holistic Data Products |
| | Expensive but reliable mainly for BI use cases with relational data on proprietary systems | Hadoop came around the 2010s, open culture soared, business use cases suffered as data reliability dropped. | Specialized stores like graph DB continue to evolve.<br><br>Focus on improving models - rapid strides in Deep Learning | Focus on data |

Figure 1.8 – Evolution of big data systems

Cloud adoption continues to grow with even highly regulated industries such as healthcare and Fintech embracing the cloud for cost-effective alternatives to keep pace with innovation; otherwise, they risk being left behind. People who have used security as the reason for not going to the cloud should be reminded that all the massive data breaches that have been splashing the media in recent years have all been from on-premises setups. Cloud architectures have more scrutiny and are in some ways more governed and secure.

## Rise of cloud data platforms

The data challenges remain the same. However, over time, the three major shifts in architecture offerings have been due to the introduction of the following:

- Data warehouses
- Hadoop heralding the start of data lakes
- Cloud data platforms refining the data lake offerings

The use cases that we've been trying to solve for all three generations can be placed into three categories, as follows:

- SQL-based BI Reporting
- **Exploratory data analytics (EDA)**
- ML

Data warehouses were good at handling modest volume structured data and excelled at BI Reporting use cases, but they had limited support for semi-structured data and practically no support for unstructured data. Their workloads could only support batch processing. Once ingested, the data was in a proprietary format, and they were expensive. So, older data would be dropped in favor of accommodating new data. Also, because they were running at capacity, interactive queries had to wait for ingestion workloads to finish to avoid putting strain on the system. There were no ML capabilities built into these systems.

Hadoop came with the promise of handling large volumes of data and could support all types of data, along with streaming capabilities. In theory, all the use cases were feasible. In practice, they weren't. *Schema on read* meant that the ingestion path was greatly simplified, and people dumped their data, but the consumption paths were more difficult. Managing the Hadoop cluster was complex, so it was a challenge to upgrade versions of software. Hive was SQL-like and was the most popular of all the Hadoop stack offerings. However, access performance was slow. So, part of the curated data was pushed into data warehouses due to their structure. This meant that data personas were left to stitch two systems that had their fair share of fragility and increased end-to-end latency.

Cloud data platforms were the next entrants who simplified the infrastructure manageability and governance aspects and delivered on the original promise of Hadoop. Extra attention was spent to prevent data lakes from turning into data swamps. The elasticity and scalability of the cloud helped contain costs and made it a worthwhile investment. Simplification efforts led to more adoption by data personas.

The following diagram summarizes the end-to-end flow of big data, along with its requirements in terms of volume, variety, and velocity. The process varies on each platform as the underlying technologies are different. The solutions have evolved from warehouses to Hadoop to cloud data platforms to help serve the three main types of use cases across different industry verticals:

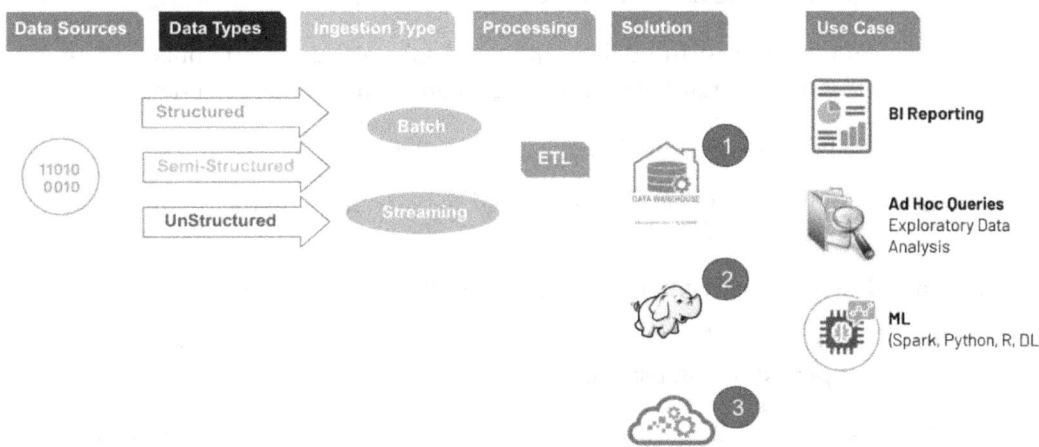

Figure 1.9 – The rise of modern cloud data platforms

# SQL and NoSQL systems

SQL databases were the forerunners before NoSQL databases arose, which were created with different semantics. There are several categories of NoSQL stores, and they can roughly be classified as follows:

- **Key-Value Stores**: For example, AWS S3, and Azure Blob Storage
- **Big Table Stores**: For example, DynamoDB, HBase, and Cassandra
- **Document Stores**: For example, CouchDB and MongoDB
- **Full Text Stores**: For example, Solr and Elastic (both based on Lucene)
- **Graph Data Stores**: For example, Neo4j
- **In-memory Data Stores**: For example, Redis and MemSQL

While relational systems honor ACID properties, NoSQL systems were designed primarily for scale and flexibility and honored BASE properties, where data consistency and integrity are not the highest concerns.

**ACID** properties are honored in a transaction, as follows:

- **Atomicity**: Either the transaction succeeds or it fails.
- **Consistency**: The logic must be correct every time.
- **Isolation**: In a multi-tenant setup with numerous operations, proper demarcation is used to avoid collisions.
- **Durability**: Once set, the data remains unchanged.

Use cases that contain highly structured data with predictable inputs and outputs, such as a financial system with a money transfer process where consistency is the main requirement.

**BASE** properties are honored, as follows:

- **Basically Available**: The system is guaranteed to be available in the event of a failure.
- **Soft State**: The state could change because of multi-node inconsistencies.
- **Eventual Consistency**: All the nodes will eventually reconcile on the last state but there may be a period of inconsistency.

This applies to less structured scenarios involving changing schemas, such as a Twitter application scanning words to determine user sentiment. High availability despite failures is the main requirement.

## OLTP and OLAP systems

It is useful to classify operational data versus analytical data. Data producers typically push data into **Operational Data Stores** (**ODS**). Previously, this data was sent to data warehouses for analytics. In recent times, the trend is changing to push the data into a data lake. Different consumers tap into the data at various stages of processing. Some may require a portion of the data from the data lake to be pushed to a data warehouse or a separate serving layer (which can be NoSQL or in-memory).

**Online Transaction Processing (OLTP)** systems are transaction-oriented, with continuous updates supporting business operations. **Online Analytical Processing (OLAP)** systems, on the other hand, are designed for decision support systems that are processing several ad hoc and complex queries to analyze the transactions and produce insights:

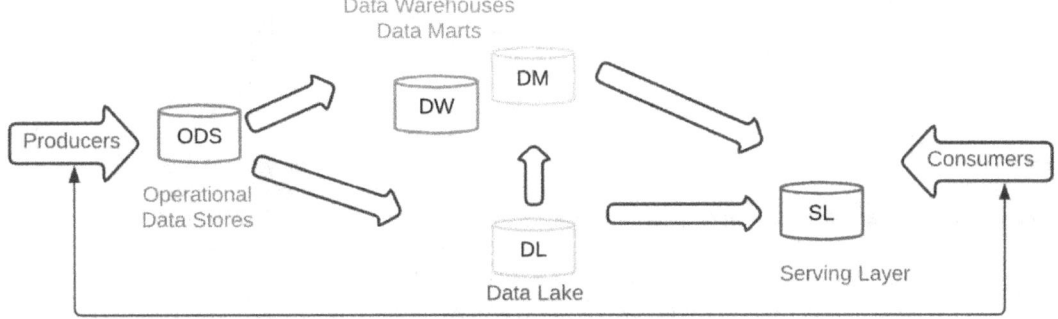

Figure 1.10 – OLTP and OLAP systems

## Data platform service models

Depending on the skill set of the data team, the timelines, the capabilities, and the flexibilities being sought, a decision needs to be made regarding the right service model. The following table summarizes the model offerings and the questions you should ask to decide on the best fit:

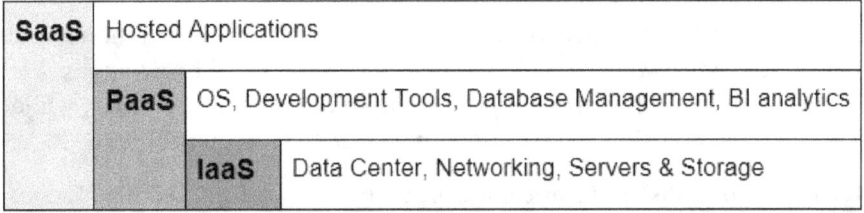

Figure 1.11 – Service model offerings

The following table further expands on the main value proposition of each service offering, highlighting the key benefits and guidelines on when to adopt them:

| | **What Is It?** | **Benefits** | **When To Use It** |
|---|---|---|---|
| **SaaS**<br>Software as<br>as a Service. | Offers third-party applications on demand over the web. | - Ease of use.<br>- Payment flexibility.<br>- Easy to customize. | When you want an app but do not have the time or resources to build or manage the software yourself. |
| **PaaS**<br>Platform as<br>a Service. | Offers a platform where a developer can design and deploy an application. | - Abstraction of computing resources.<br>- Full control of the features and tools.<br>- Seamless platform updates. | When you want to develop and customize your application without worrying about the infrastructure. |
| **IaaS**<br>Infrastructure<br>as a Service. | Offers the necessary cloud computing infrastructure to perform generalized or specialized tasks. | - Dynamic scaling.<br>- Save money by only paying for what you use. | When you want just the cloud-based computing resources and not any software. |

Figure 1.12 – How to find the right service model fit

# Distributed computing

Scalability refers to a system's ability to adapt to an increase in load without degrading performance. There are two ways to scale a system – vertically and horizontally. Vertical scaling refers to using a bigger instance type with more compute horsepower, while horizontal scaling refers to using more of the same node type to distribute the load.

In general terms, a process is an instance of a program that's being executed. It consists of several activities and each activity is a series of tasks. In the big data space, there is a lot of data to crunch, so there's a need to improve computing speeds by increasing the level of **parallelization**. There are several multiprocessor architectures, and it is important to understand the nuances to pick linearly scalable architectures that can not only accommodate present volumes but also future increases.

# SMP and MPP computing

Both **symmetric multi-processing** (**SMP**) and MPP are multiprocessor systems.

As data volume grows, SMP architectures transition to MPP ones. MPP is designed to handle multiple operations simultaneously by several processing units. Each processing unit works independently with its resources, including its operating system and dedicated memory. Let's take a closer look:

- **SMP**: All the processing units share the same resources (operating system, memory, and disk storage) and are connected on a system bus. This becomes the choke factor of the architectures scaling linearly:

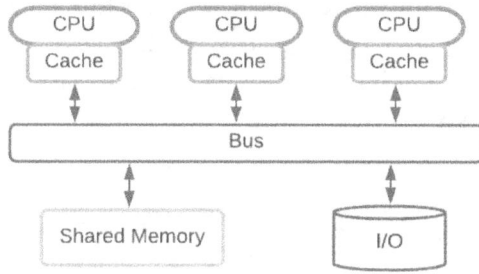

Figure 1.13 – SMP

- **MPP**: Each processor has its own set of resources and is fully independent and isolated from other processors. Examples of popular MPP databases include Teradata, GreenPlum, Vertica, AWS Redshift, and many more:

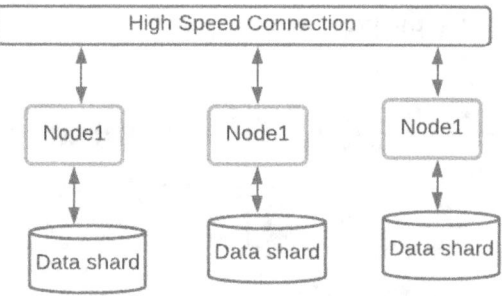

Figure 1.14 – MPP

In the next section, we'll explore Hadoop and Spark, which are newer entrants to the space, and the map/reduce and **Resilient Distributed Datasets** (**RDDs**) concepts, which mimic the parallelism constructs of MPP databases.

# Parallel and distributed computing

Advances in distributed computing have pushed the envelope on compute speeds and made this process possible. It is important to note that *parallel processing is a type of distributed processing.* Let's take a closer look:

- **Parallel Processing**:

  In parallel processing, all the processors have access to a single shared memory (https://en.wikipedia.org/wiki/Shared_memory_ architecture) instead of having to exchange information by passing messages between the processors:

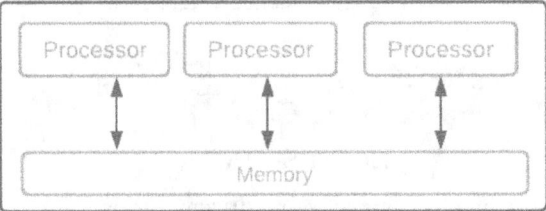

Figure 1.15 – Parallel processing

- **Distributed Processing**:

  In distributed processing, the processors have access to their own memory pool:

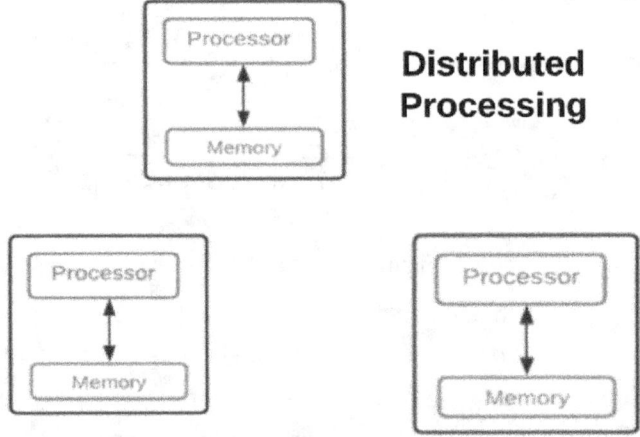

Figure 1.16 – Distributed computing

The two most popular distributed architectures are Hadoop and Spark. Let's look at them in more detail.

## Hadoop

Hadoop is an Apache open source project that started as a Yahoo! project in 2006. It promises to provide an inexpensive, reliable, and scalable framework. Several distributions, such as Cloudera, Hortonworks, MapR, and EMR, have offered packaging variations. It is compatible with many types of hardware where it runs as an appliance. It works with scalable distributed filesystems such as S3, HFTP FS, and HDFS with multiple replications on commodity-grade hardware and has a service-oriented architecture with many open source components.

It has a master-slave architecture that follows the map/reduce model. The three main components of the Hadoop framework are HDFS for storage, YARN for resource management, and Map Reduce as the application layer. The HDFS data is broken into blocks, replicated a certain number of times, and sent to worker nodes where they are processed in parallel. It consists of a series of map and reduce jobs. NameNode keeps track of everything in the cluster. As the resource manager, YARN allocates the resources in a multi-tenant environment. JobTracker and TaskTracker monitor the progress of a job. All the results from the MapReduce stage are then aggregated and written back to disk in HDFS:

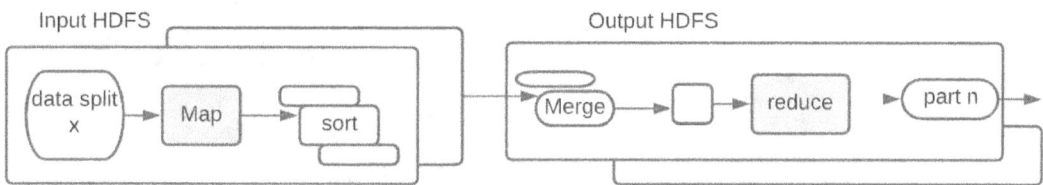

Figure 1.17 – Hadoop map/reduce architecture

## Spark

**Spark** is an Apache open source project that started in 2012, at AMPLab (`https://amplab.cs.berkeley.edu/`) at UC Berkeley. It was written in Scala and provides support for the Scala, Java, Python, R, and SQL languages. It has connectors for several disparate providers/consumers. In Spark lingo, a job is broken into several stages and each stage is broken into several tasks that are executed by executors on cores. Data is broken into partitions that are processed in parallel on worker node cores. So, being able to partition effectively and having sufficient cores is what enables Spark to be horizontally scalable:

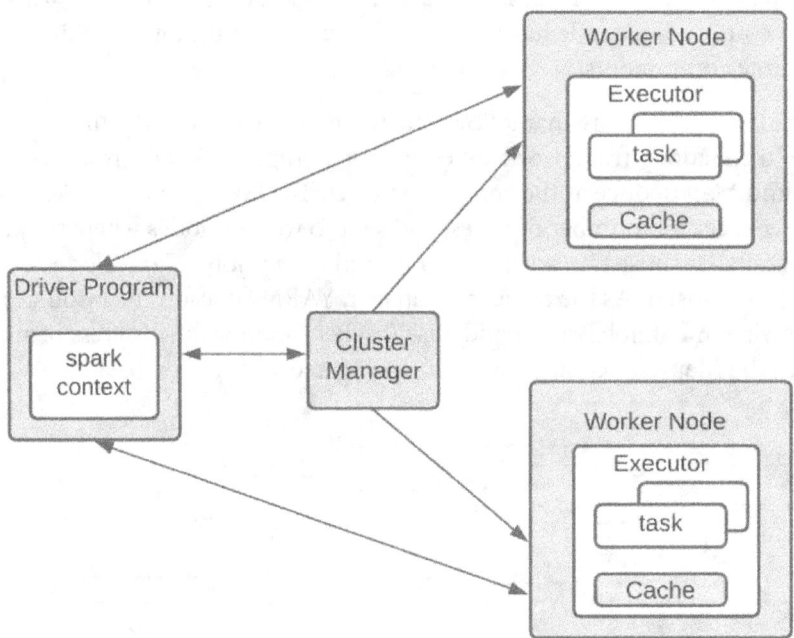

Figure 1.18 – Spark distributed computing architecture

Spark is a favorite tool in the world of big data, not only for its speed but also its multifaceted capabilities. This makes it favorable for a wide variety of data personas working on a wide range of use cases. It is no wonder that it is regarded as a Swiss Army knife for data processing:

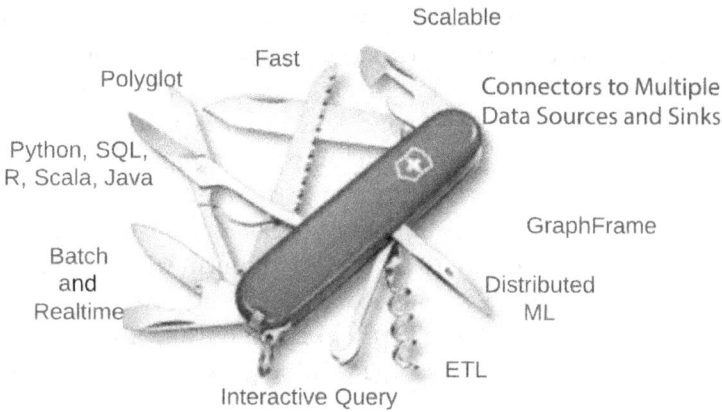

Scalable

Fast

Polyglot

Connectors to Multiple
Data Sources and Sinks

Python, SQL,
R, Scala, Java

GraphFrame

Batch
and
Realtime

Distributed
ML

ETL

Interactive Query

Figure 1.19 – Spark is a Swiss Army knife in the world of data

## Hadoop versus Spark

Spark is *~100x* faster in-memory than Hadoop. This is on account of more disk operations in Hadoop, where each map and reduce operation in a job chain goes to disk. Spark, on the other hand, processes and retains data in memory for subsequent steps in a **Directed Acyclic Graph** (**DAG**). Spark processes data in RAM using a concept known as a **Resilient Distributed Dataset** (**RDD**), which is immutable. So, every *transformation* is a node in the DAG that is lazily evaluated when it encounters an explicit *action*. Although Spark is a standalone technology, it was also packaged with the Hadoop ecosystem to provide an alternative to Map Reduce. Hadoop is losing favor and is on the decline, whereas Spark continues to be an industry favorite.

# Business justification for tech spending

Tech enthusiasts with their love for bleeding-edge tools sometimes forget why they are building a data product. Research and exploration are important for innovation, but it needs to be disciplined and controlled. Not keeping the business counterparts in the loop results in miscommunication and misunderstandings regarding where the effort is going. Ego battles hinder project progress and result in wasted money, time, and people resources, which hurts the business. Tech should always add value and growth to a business rather than being viewed as a cost allocation. So, it is important to demonstrate the value of tech investment.

A joint business-technology strategy helps clarify the role of technology in driving business value to provide a transformation agenda. **Key performance indicators** (**KPIs**) and metrics including growth, **return on investment** (**ROI**), profitability, market share, earnings per share, margins, and revenue help quantify this investment.

The execution time of these projects is usually significant, so it is important to achieve the end goal in an agile manner in well-articulated baby steps. Some of the benefits may not be immediately realized, so it is important to balance infrastructure gains with productivity and capability gains and consider **capital expenditure on initial infrastructure investment (CAPEX)** versus **ongoing operating expenses (OPEX)** over a certain period. In addition, it is always good to do frequent risk assessments and have backup plans. Despite the best projections, costs can escalate to uncomfortable and unpredictable heights, so it is important to invest in a platform with tunable costs so that it can easily be monitored and adjusted when needed. Data is an asset and must be governed and protected from inappropriate access or breaches. Not only are such threats expensive, but they also damage the reputation of the organization:

|  | Technology | Business |
|---|---|---|
| **Present** | Current Technology Challenges | Negative Business Consequences |
| **Future** | Proposed Technology Changes | Positive Business Outcomes |

Figure 1.20 – Mapping the impact of technology on business outcomes

The preceding table captures the current technological challenges and their subsequent negative business consequences. Articulating the proposed tech changes and their potential positive business outcomes, helps make the case for the transformation

# Strategy for business transformation to use data as an asset

Data-driven organizations exhibit a culture of analytics. This cannot be confined to just a few premiere groups but rather to the entire organization. There are both cultural and technical challenges to overcome and this is where people, processes, and tools need to come together to bring around sustainable changes. Every business needs a strategy for business transformation. Here are some best practices for managing a big data initiative:

- Understand the objectives and goals to come up with an overall enterprise strategy.

- Assess the current state and document all the use cases and data sources.

- Develop a roadmap that can be shared for collaborating and deciding on the main tools and frameworks to leverage organization-wide.

- Design for data democratization to allow people to have access to data they have access to.

- Establish rules around data governance so that workflows can be automated correctly without fear of data exfiltration.

- Manage change as a continuous cycle of improvement. This means that there should be a center of excellence team that can serve as a hub and spoke model that interfaces with the individual lines of business. Adequate emphasis should be placed on training to engage and educate the team.

# Big data trends and best practices

*"The old order changeth, yielding place to new..."*

*(The Passing of Arthur, Alfred Lord Tennyson, 1809–1892)*

We are living in an age of fast innovation and technology changes that are happening in the blink of an eye. We can learn from history and learn from the mistakes of those before us. However, we don't have the luxury to analyze everything around us and understand the top trends, though this will give us a better appreciation of the landscape and help us gravitate toward the right technology for our needs.

There is an increase in the adoption of cloud infrastructure because of the following points:

- It provides affordable and scalable storage.

- It's an elastic distributed compute infrastructure with pay-as-you-go flexibility.

- It's a multi-cloud strategy and some on-premises presence to hedge risks.

- It provides an increase in data consolidation to break down individual data silos in data lakes.

- Other data stores such as data warehouses continue to live on, while newer ones such as lakehouses and data meshes are being introduced.

- Unstructured data usage is on the rise.

- Improved speed to insights.

- Convergence of big data and ML.

- Detecting and responding to pattern signals in real time as opposed to batch.

- Analytics has moved from simple BI reporting to ML and AI as industries move from descriptive analytics to prescriptive and finally predictive.

- Improved governance and security

- Data discovery using business and operational enterprise-level meta stores.

- Data governance to control who has access to what data.

- Data lineage and data quality to determine how reliable the data is.

Let's summarize some of the best practices for building robust and reliable data platforms:

- Build decoupled (storage and compute) systems because storage is far cheaper than compute. So, the ability to turn off compute when it's not in use will be a big cost saving. Having a microservices architecture will help manage such changes.

- Leverage cloud storage, preferably in an open format.

- Use the right tool for the right job.

- Break down the data silos and create a single view of the data so that multiple use cases can leverage the same data with different tools.

- Design data solutions with due consideration to use case-specific trade-offs such as latency, throughput, and access patterns.

- Log design patterns where you maintain immutable logs for audit, compliance, and traceability requirements.

- Expose multiple views of the data for consumers with different access privileges instead of copying the datasets multiple times to make slight changes to the data access requirements.

- There will always be a point where a team will have to decide between whether they build or buy. Speed to insights should guide this decision, irrespective of how smart the team is or whether there is a window of opportunity, and you should not lose it in the pursuit of tech pleasures. The cost of building a solution to cater to an immediate need should be compared with the cost of a missed opportunity.

# Summary

In this chapter, we covered the role of data engineering in building data products to solve a host of use cases across diverse industry verticals. A whole village of data personas come together to create a sound robust data application that provides valuable insights for business. We briefly looked at the big data landscape and the metamorphosis it has gone through over the years to arrive at present-day modern cloud data platforms. We talked about the various distributed architectures that we can use to crunch data at scale. We emphasized that only by pulling business and tech together can we create a symbiotic and data-driven culture that spurs innovation to put a company ahead of its competitors.

In the next chapter, we will explore data modeling and data formats so that your storage and retrieval operations are optimized for the use case at hand.

# 2

# Data Modeling and ETL

*"Poor programmers care about code, and good programmers care about the data structure and the relationships between data."*

*— Linus Torvalds, the founder of Linux, on the importance of data modeling*

In the previous chapter, we introduced the big data ecosystem, the use cases across different industry verticals that use this data, the common challenges that they all face, and their journey towards digitization. We also looked at the trends in compute and storage technologies along with cloud adoption that is paving the way to enable companies to be more data-driven.

Data platforms are continuously evolving to support business analytic use cases and speed to insights is critical for a business to remain relevant and competitive. Both BI and AI leverage curated data to produce sound insights, but getting to curated data requires some discipline around data layout, modeling, and governance. In this chapter, we will look at ways to design and build robust pipelines to host the data by first tackling data modeling concerns, and then moving on to metadata management to make the data easily discoverable. We will compare different data formats and compression techniques for operational efficiencies and look at common design patterns and best practices for designing these pipelines.

In particular, we will be covering the following topics in this chapter:

- What is data modeling and why should you care?
- Understanding metadata – data about data
- How to choose the right data format
- Storage is cheap so why bother with data compression?
- ETL using structured/semi-structured data and best practices
- Common big data design patterns

# Technical requirements

To follow along this chapter, make sure you have the code and instructions as detailed in this GitHub location: `https://github.com/PacktPublishing/Simplifying-Data-Engineering-and-Analytics-with-Delta/tree/main/Chapter02`

Let's get started!

# What is data modeling and why should you care?

When you design software, you start architecting with a paper design of the various components and how they interact. The same is true of big data systems. To get the most value out of data, you need to understand its intrinsic properties and inherent relationships. Data modeling is the process of organizing, representing, and visualizing your data so it fits the needs of the business processes. There are several functional and nonfunctional requirements around data, technology, and business that should be taken into consideration. Business operational processes and the structure of the generated data from the various operations are inputs to the data model. Let's look at some of the advantages of going through data modeling before rushing to implement a data solution.

## Advantages of a data modeling exercise

At its core, data modeling is designed for persisting data and retrieving it in an optimal way. The following lists some of the main benefits of not skipping this step:

- The visual representation of relationships helps provide a common mental image of the solution and prevents incorrect interpretations.

- Defining the data constraints explicitly helps enforce rules to improve data quality.

- The paper exercise also increases consistency in naming conventions, rule definitions, overall security, and semantic aspects while improving the core data analytics.

- It also provides common vocabulary not only to the technical folks but also to the business counterparts, and this clear definition is exposed in catalogs leading to better discovery and faster metadata searches by other parts of the organization.

- Inconsistencies are discovered early on as part of the review process, which results in cheaper fixes.

There are different stages in data modeling, including conceptual, logical, and physical stages and in the next section, we will look at the cascading nature of this design.

## Stages of data modeling

There are three main stages of data modeling, as follows:

- **Conceptual data model**: This is the simplest visual representation of entities and their relations and is meant for general consumption, but mainly for the business stakeholders. It articulates the basic schema, rules of ingestion, and consumption, and because of its simplicity, it serves as a way to bridge the gap between the different data personas and align them on the same page.

  Let's model a retail scenario where a registered customer browses through a catalog of products to make a purchase. Although there are several entities involved here, for simplicity's sake, let's consider the **customer** and the **product** as the primary entities and the **transaction** as the primary event that needs to be captured. To bill the customer, we need to capture some profile information, including the customer's address. Nouns are typically used to capture entities and verbs to capture the operations or interactions between the entities. In this case, *customer* **lives at** an *address* and makes a purchase/sale of a given *product* from the catalog. Each entity has attributes, as do the interactions, and these details are captured in a conceptual model, as shown in the following diagram:

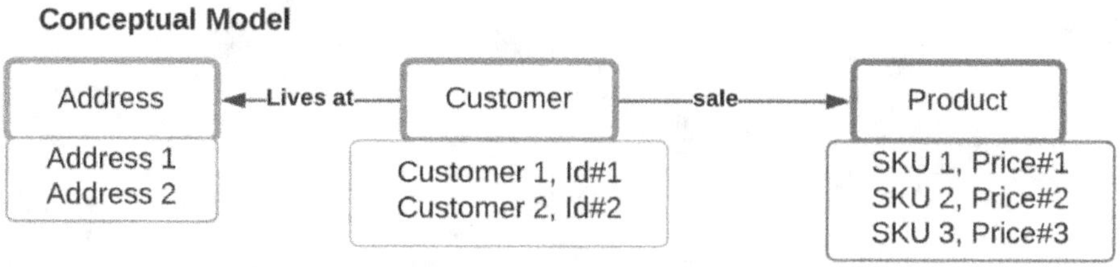

Figure 2.1 – Conceptual model

- **Logical data model**: This builds on the conceptual model and further refines the structure with additional details. This is usually done by a data architect and does not make assumptions about the exact implementation.

The entities and relationships were captured as part of the conceptual model; they are further examined to articulate data types, associations, and cardinality. For example, the customer ID is a numeric data type and the customer name is a string data type; moreover, the address entity is very tightly associated with a customer. A customer is typically associated with a single billing address but has a lot of past product purchase history. So, the cardinality between **Customer** and **Address** is a **1:1** relation, whereas with **Product**, it is a 1:many relation. The following diagram captures the additional refinement details.

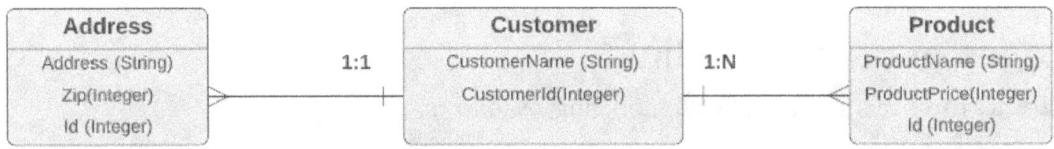

Figure 2.2 – Logical model

- **Physical data model**: This is the final stage where data-store-specific details around data types, constraints, and other relationships are ironed out, usually by a database modeler.

Taking the same retail example, the physical model design adds platform-specific implementation details. For example, the customer ID is the primary key to identifying the customer uniquely; the numeric type has been identified to be an **Integer** field, as shown in the following diagram:

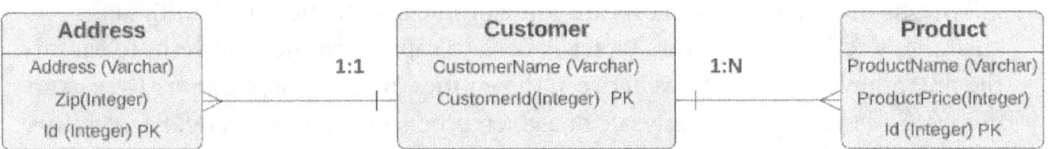

Figure 2.3 – Physical model

To summarize, data modeling starts with a conceptual view of the data and progresses to a logical and, finally, a detailed physical model. This is subject to change because existing datasets evolve and newer data sources continue to be added requiring the relationships to be redrawn, which is why it is an iterative process. In the next section, we will look at techniques for modeling different data stores; for example, an in-memory store and a graph data store will require different layouts.

# Data modeling approaches for different data stores

Data modeling techniques depend on the underlying data store. Regardless of the system, it should include business requirements, governance and security mandates, physical storage, integration interfaces, and the ability to handle the various data types. For example, in the context of the earlier example, this is the time to identify the key consumption patterns around key entries such as location, date, product **Stock Keeping Unit (SKU)**, and the like.

The focus should be on designing a system and not a schema, as a schema evolves over time. Also, it is easy to get overwhelmed with all the datasets in the ecosystem, so you should focus on the core data that is critical to your business and get that modeled correctly first. Care should be taken to improve the quality of metadata, as better discoverability allows it to be placed appropriately into the data models to support business. Since the world of data is divided between relational and non-relational, let's look at modeling approaches for each.

NoSQL data stores encourage denormalization and wide tables to improve read performance by adding redundant copies of the data to avoid joins. Graph data stores use nodes and edges and define properties on both nodes and edges to model the key entities and their relationships. In-memory data stores use key-value stores for fast data retrieval.

## Relational data modeling

Normalizations, usually in **third normal form (3NF)**, to reduce data redundancies and improve data integrity are standard practices when designing SQL data stores. A star or snowflake schema is usually adopted in designing data marts.

**Online transaction processing (OLTP)** systems and **online analytical processing (OLAP)** systems have different requirements. One is operationally focused, and the other is decision support and analytics-focused. So, the design of these systems is very different. **Entity relation (ER)** modeling is used for OLTP systems and **dimensional modeling** for OLAP systems. OLAP systems consist of facts (transactions) at the center and are surrounded by dimensions (attributes) tables where facts change frequently, and dimensions change slowly. Dimension modeling is further categorized by **star** and **snowflake** schemas, as shown in the following figure:

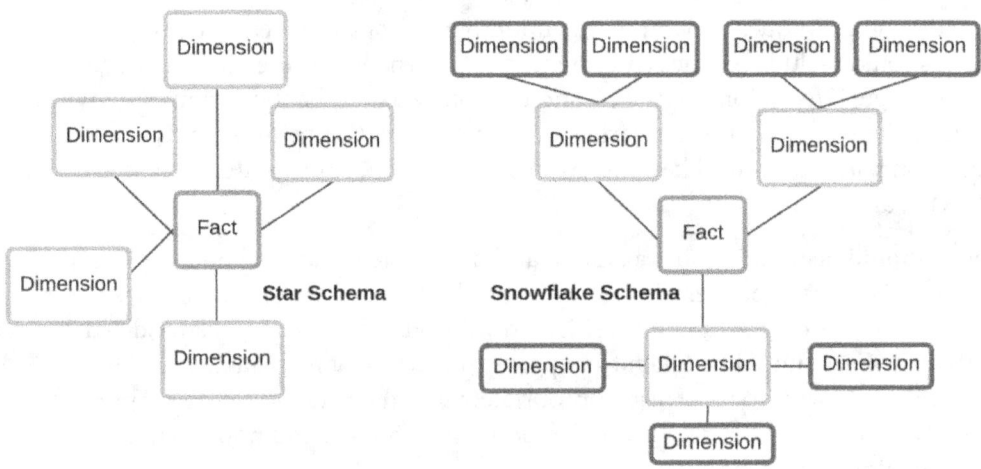

Figure 2.4 – Star schema versus snowflake schema

Star is simpler with a central fact table and a single layer of surrounding dimension tables and is more widely adopted in the industry as it is easier to maintain, whereas snowflake is more involved with multiple layers of dimension tables, allowing them to be truly normalized.

Bill Inmon and Ralph Kimball are considered the fathers of modern data warehouse and data mart design. They regarded the data warehouse as the source of truth for all business reporting. However, the implementation of their approaches is different.

Bill Inmon advocated a top-down approach, where you start from the data warehouse and break it down into specialized data marts on a need basis to meet the requirements of various departments within the organization, such as a finance mart or an accounting mart, an HR mart, and other similar functions. Although it is more structured and easier to maintain, it is rigid and takes more time to build.

Ralph Kimball advocated a bottom-up approach, where the dimensional data marts are first created to provide reporting and analytical capabilities for specific lines of business before creating the warehouse itself. It is considered to be more scalable, and the **return on investment (ROI)** is usually realized faster, although reuse is low.

## Non-relational data modeling

Since NoSQL is schema-free, it leads people to believe that you don't need a data model with NoSQL technologies. That is not correct. You do have to define the various dimensions of how you plan to organize your data.

There are several flavors of non-relational data stores, such as the following:

- **Key-value**: S3 and Azure Blob storage
- **Bigtable**: DynamoDB, HBase, and Cassandra
- **Document**: Mongo and CouchDB
- **Full-text search**: Elastic and Solr
- **Graph**: Neo4j, Amazon Neptune, and ArangoDB

In relational systems, we have the data and that drives what kind of questions or queries it can support. In non-relational systems, we start with the consumption patterns and design the store and keys accordingly.

Data duplication and denormalization are the accepted norm. Joins are less prevalent as joins are more expensive and sometimes not supported at all by the underlying data store. Wide tables are a norm and, unlike a **Relational Database Management System (RDBMS)**, the maximum number of columns is not a hard number. Data can be linked either by embedding (keeping data that is frequently used together) or by referencing using unique keys (identifying and modeling the relationships), as shown in the following snippet:

```
        Embedded                         Referenced
                                  {
  {                                 "order_ID":"0001",
    "order_ID":"0001",            "Product":"0002"
                                  }
    "product":{
      "product_name":"Pencil",    { "product_ID":"0002",
      "product_category":"supplies",  "product_name":"Pencil",
      "list_price":"10.00"          "product_category":"supplies",
    }                               "list_price":"10.00"
  }                               }
```

Figure 2.5 – Embedding versus referencing data across tables

Aggregates are handled using either a **composite key** or **column families**. A composite row key includes multiple data elements, which can be useful for grouping rows together or for finding a range of keys. In a distributed setup, we want all worker nodes to be well-utilized. If all keys are similar, that leads to **hotspotting**, which puts undue pressure on a few nodes and so should be avoided. For example, if a timestamp is part of the key, then reversing it helps to distribute it better. All the newly created data in that batch does not get delegated to one node while the others are underutilized. Another strategy is to add a hash prefix to the natural row key in order to improve the key distribution, and still have an additional grouping mechanism.

Column families are used in data stores such as Cassandra, HBase, and DynamoDB, which are columnar databases. Related fields can further be grouped in column families. Another characteristic is that unlike a table in a relational database, the different rows in the same table or column family do not have to share the same set of columns. For example, you can have one large product catalog table of diverse families of products that may share some common attributes, such as SKU, color, and price, but also have a lot of other unique attributes for each product. For example, the attributes to define a brush will be different from the attributes to define a paint, but they are both placed in the same table. This leads to a lot of unfilled columns. The advantage is that the lookup is simplified and joins against separate feature tables are eliminated.

**Document stores** such as MongoDB use JSON data. **Search indices** can use XML (such as Solr) or JSON (such as Elasticsearch). **Graph databases** use nodes and edges, both of which have properties, and relationships between the nodes are captured as either directed or undirected edges.

A **tree/adjacency list** is another technique. Relational databases are not adequate for representing hierarchical or graph-like data modeling and processing. NoSQL solutions offer better solutions. For example, parent children's relationships can be flattened and captured, as shown in the following example:

```
{    "id": "123",    "type":"country",    "children": ["State1",
"State2"],   "parent": null }
{    "id": "456",      "type":"state",    "children": ["Zip1", "Zip2",
"Zip3"],    "parent": "123" }
```

The graphical representation of the involved tiered relationship is shown here:

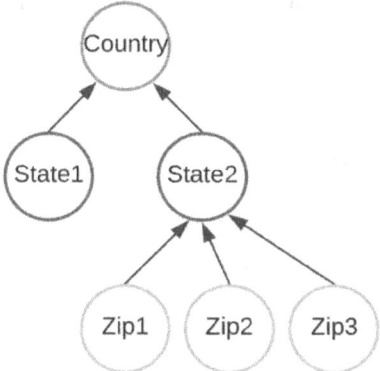

Figure 2.6 – Hierarchical data modeling

In this section, we examined different techniques to model big data. In the next section, we will look at what metadata is and why it is almost as important, if not more important, than the data itself.

# Understanding metadata – data about data

Metadata is data about the data and is an important governance aspect exposed in data catalogs. The data value use case is around the ability to identify key data assets and assess their economic importance to the organization. Let's examine different aspects of metadata.

## Data catalog

A **catalog** is a tool that houses the metadata and provides the tooling for search and discoverability. This is often confused with data dictionaries, which are just data artifacts and do not necessarily have the associated tooling to facilitate data search and retrieval.

There are several vendors in this space and some of the popular ones include Collibra, Alation, and Glue. The data discovery use case is probably the most valuable as it helps users (data engineers, data analysts, and data scientists) search, find, and understand data.

**Data governance** is another important capability, where data lineage is documented in a central place and data freshness can be ascertained for auditing and compliance purposes. From a data privacy and risk perspective, it helps to discover and document sensitive data, such as **Personally Identifiable Information** (**PII**) and **Protected Health Information** (**PHI**), and its flow through the data landscape.

# Types of metadata

Metadata can be classified into three categories, as illustrated in the following diagram:

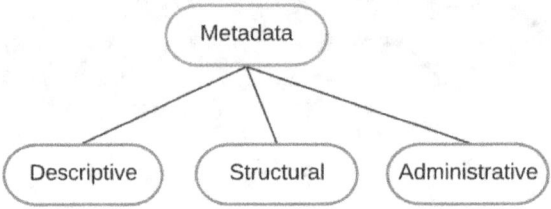

Figure 2.7 – Types of metadata

The types of metadata are as described here:

- Descriptive metadata includes definitions, ownership details, and general content.

- Structural metadata refers to the organization of the data and its relationship to other datasets.

- Administrative metadata includes the dos and don'ts around how to add/change and who can access it.

This is mainly stored in three types of meta stores, as shown in the following figure:

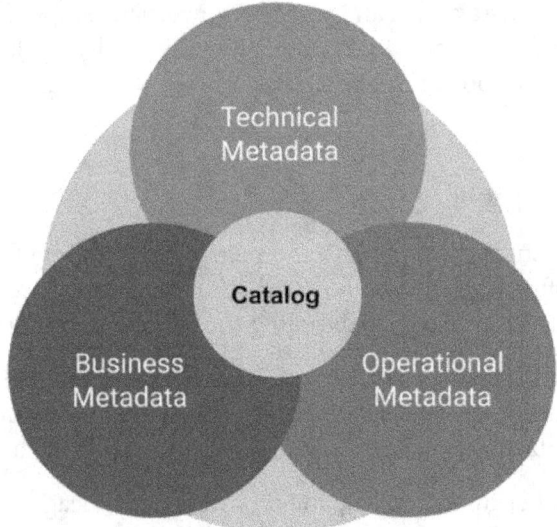

Figure 2.8 – Types of meta stores

Each of these meta stores addresses a different dimension and together they comprise the overall catalog:

- **Technical metadata**: This gives information on the format and structure of the data. This refers to schema and data models, data lineage, and access control lists, which determine permission levels of who has privileges to use the data.

- **Business metadata**: This defines common business terms, such as table and column definitions, so it can be consumed by both technical and non-technical users. It articulates domain and business-specific rules, data quality, and sharing rules.

- **Operational metadata**: This defines the data life cycle of when current data is archived and purged, along with data lineage of where it comes from and how it has been transformed to provide transparent audit trails.

It goes without saying that it is good to have these additional nuggets of information to understand the data better. In the next section, we will go a step further and justify why management of the metadata is crucial to business operations, which is why it is termed *the nerve center of data*.

# Why is metadata management the nerve center of data?

Metadata is critical in driving business value. It does this by facilitating innovation and collaboration among data teams, which indirectly helps mitigate risks such as misinterpretation and misrepresentation of data. Not only does it help ML practitioners discover the right datasets to use for their modeling exercises, but it also enables citizen data scientists to access the most valuable datasets, thereby ensuring the generation of timely and accurate insights.

There has been an increase in the rate of adoption of metadata management systems on account of an increase in the need for data governance, regulatory and compliance requirements, and an increase in self-service capabilities to support data democratization. There is evidence to prove that a data-centric approach to improving insights is more effective than a model-centric approach (that is, algorithms), meaning that the effort spent in producing *good data* is well worth it, and having a centrally governed catalog as a source of truth helps achieve it. In recent years, non-traditional data sources such as social media have contributed to large volumes of complex unstructured data, and more business users are now leveraging this data to make critical business decisions. Businesses all around are modernizing with digital initiatives requiring a single source of trusted data to be the bloodline for everyday operations.

The *What is metadata management?* article (`https://www.collibra.com/us/en/blog/what-is-metadata-management`) summarizes the catalog as follows:

> *"a cross-organizational agreement on how to define informational assets for converting data into an enterprise asset. As data volumes and diversity grow, metadata management is even more critical to derive business value from the gigantic amounts of data."*

A good catalog helps address questions such as *who/what/where/how* in the following context:

- **Created the data**: This is where the business definitions are articulated, including schema definition and storage location.

- **Owns the data**: Here, the life cycle requirements are laid out.

- **Data steward regulating changes to the dataset**: This includes several governance criteria, including privacy and security policies around data access

- **Consumers of the data**: The business drivers for data usage and sharing along with security and privacy requirements.

We will now move on to understanding more about how to move and transform data using ETL.

# Moving and transforming data using ETL

A data pipeline is an artifact of a data engineering process. It transforms raw data into data ready for analytics. These in turn help solve problems, aid support decisions, and make our lives more convenient. In some ways, it can be thought of as the stitch between the OLTP and OLAP systems. Data pipelines are sometimes referred to as **ETL**, which stands for **extract, transform, load**, and it has a variation called **extract, load, transform** (**ELT**). The main difference between the two is whether the incoming data is first saved to disk and then transformed (data wrangling) or vice versa. The processing is loosely referred to as ETL. Although, it is fair to say ELT is relevant in the context of Data Lakes and unstructured data, whereas ETL is used for Data Warehouses. The following diagram shows how ETL bridges the gap between the OLTP and OLAP systems:

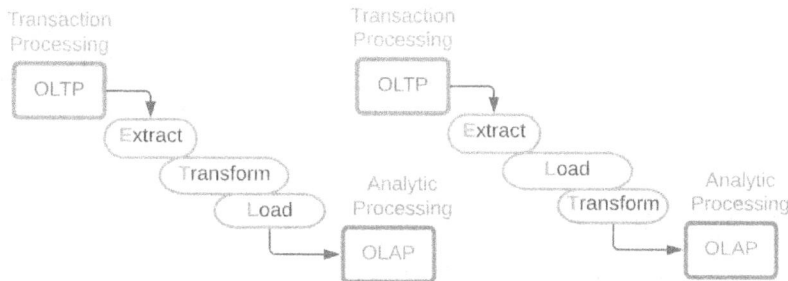

Figure 2.9 – ETL stitches OLTP and OLAP systems

Data pipelines include a set of processes to design, test, deploy, monitor data flow, handle changes, and redeploy as necessary and a lot of automation to bring some discipline to the ever-changing data landscape. The main benefits are to support real-time analytics and applications that drive real-time decisions, reduce dependency on IT folk by making data available via self-serve channels, and help accelerate cloud adoption with dynamic resources.

The main challenges include relentlessly chasing data issues that include schema and quality changes (data drift). Sometimes, fixing these issues can cause outages and delays to existing jobs. This is tied tightly to the underlying infrastructure, process, and technology and can be vulnerable to any changes there. For example, a temporary glitch in the cloud ecosystem will result in a failure of the data pipeline. Let's look at the different scenarios for designing these ETL pipelines.

# Scenarios to consider for building ETL pipelines

There are several ways to categorize pipelines. The first level of classification is batch and streaming workloads. The most common scenarios include periodic (batch and micro-batch) and continuous ingestion, bulk migration, change data capture, and slowly changing dimensions, which we will look at in the following sections.

## Periodic and continuous ingestion

This refers to batch/micro-batch and pure streaming ingestion. As the name suggests, one deals with bounded data and the other with unbounded or continuously flowing data. In some ways, a streaming pipeline is a batch pipeline that is triggered at a certain frequency that is large enough to make it seem like a well-defined bounded dataset. So once a month, once a week, daily, or hourly all make for good batch workloads. Streaming is of two types, micro-batch and continuous. Micro-batch, as the name suggests, is still batch-like with very low frequency, typically in milliseconds, and continuous means that as soon as the data arrives, it is processed; there is no buffering.

## Bulk data migration

This is a one-time bulk load of historical data between two systems. Usually, this is done infrequently and has to be planned very carefully to avoid excessive downtime as the cutover is planned.

## Change data capture

Keeping two disparate data stores in sync is another challenge, and this is where **change data capture** (**CDC**) comes in handy. The changes in the source system are passed as change data feed to the target system. For example, each row is identified by an operation (insert, update, and delete) along with the new values so the target system can decide how to merge these values. The following diagram shows the flow of this data from the original data producer OLTP system all the way to the consumer via the OLAP system:

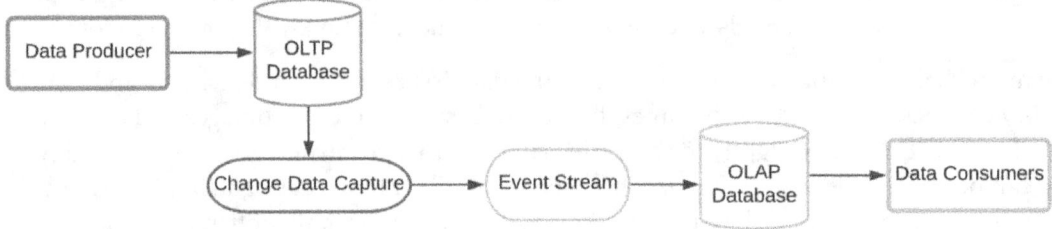

Figure 2.10 – CDC from the perspective of changes to the source system

## Slowly changing dimensions

**Slowly Changing Dimensions** (**SCD**) is similar in concept to CDC except that it is focused on the destination target tables. An OLAP system is built with central fact tables and surrounding dimension tables. It is the facts that change frequently, and dimensions change rather infrequently, hence the term *slowly changing dimensions*. Different use cases have different needs on how the change is to be managed, so there are a few variations, and we will cover them in *Chapter 6, Solving Common Data Pattern Scenarios with Delta*.

# Job orchestration

This is part of the automation process and refers to scheduling the pipeline either via cron or programmatically using an API call to trigger a pipeline. In any case, a pipeline should be smart, meaning it should know how to auto-heal, notify on errors, and retry when necessary.

In this section, we looked into how data is ingested and transformed either one time or periodically from various operational stores into a central data store for the purposes of analysis. In the next section, we will explore the various considerations while choosing the right data format for efficiencies in data storage and retrieval of different use cases.

# How to choose the right data format

Not all tools support all of the data formats. Every tool reads data off disk in chunks of blocks (KB/MB/GB). Minimizing these fetches helps improve the speed of access to data. Conversely, a single read for a single record brings back a lot more data than you may want, so caching it may help with subsequent queries. Different systems have different default block sizes. To choose the right data format, you need to consider several factors, such as the following:

- What is the optimal tradeoff between cost, performance, and throughput considerations of ingestion and access patterns?

- Are you constrained by storage or memory or CPU or I/O?

- How large is a file? If your data is not splitable, we lose the parallelism that allows fast queries.

- How many columns are being stored, and how many columns are used for the analysis?

- Does your data change over time? If it does, how often does it happen, and how does it change?

Let's examine different file formats to compare and contrast their main characteristics, which will help to pick the right one for the use case under consideration.

## Text format versus binary format

The data is encoded differently; in text, the data is stored as readable characters, whereas in binary format, it is stored as a sequence of bytes. Examples of text format are CSV and JSON, which are human-readable, whereas examples of binary format include ORC, Parquet, and Avro.

## Row versus column formats

In this section, we will examine the differences between row and column data formats in the context of an example table with 1,000 columns and a million rows and a few typical data patterns on this table, such as the following list, which includes typical data operations of selecting, inserting, and updating data:

1. Insert a row of this table.
2. Update a few columns.
3. Select five columns for a given partition (logical grouping of data).
4. Select all fields for a few given keys.

Examples of row-oriented data formats include CSV, JSON, and Avro, and are best suited for OLTP systems where data is stored and retrieved an entire row at a time. So, use cases 1 and 2 would be good candidates for such a file format. If a user issues a select statement and specifies a few columns, it is not much different from selecting all the columns. In other words, use cases 3 and 4 will require similar resources. Because the retrieval strategy is to get the full row, this could read unnecessary data if only some of the data in that row is required. The typical compression mechanisms in row-oriented formats are less effective as compared to column-oriented ones.

Examples of column-oriented data formats include ORC and Parquet and are best suited for OLAP systems where data is stored and retrieved in columns. Use case 2 from the previous list is an ideal case because only a few relevant columns are touched. Unlike row-oriented format, it is able to read only the relevant data if required, thereby conserving space in block chunks transferred back and forth from disk to memory. Read and write operations are typically slower though, as compared to row-oriented. These are more efficient in performing operations applicable to the entire dataset, such as aggregations over many rows and columns. Higher compression rates are possible due to fewer distinct values in columns.

Let's review some of the properties to consider while deciding on a file format to best support the use case at hand.

## When to use which format

With all the choices out there, we need some guidance on which format to choose. The following table lists the various preferred features and the top popular formats as columns. The more green, the better. Parquet and Avro are the forerunners:

| Properties | CSV | JSON | Parquet | Avro |
|---|---|---|---|---|
| Columnar | ✗ | ✗ | ✓ | ✗ |
| Compressible | ✓ | ✓ | ✓ | ✓ |
| Splittable | ✓ | ✓ | ✓ | ✓ |
| Readable | ✓ | ✓ | ✗ | ✗ |
| Complex Data | ✗ | ✓ | ✓ | ✓ |
| Schema | ✗ | ✗ | ✓ | ✓ |

Figure 2.11 – Comparing data formats

The two main considerations for choosing a suitable file format are with regard to data storage, level of normalization, and query patterns.

Row stores have the ability to write data quickly because the whole row is stored together as a single read/write, and so, are good for transaction processing. The same is true of updates, as normalization of data makes updates more efficient in a row store.

Column stores are good at aggregating large volumes of data for a subset of columns. In big data, column-based data formats are preferred for analytic querying because of better compression and performance. Columnar data stores prefer a denormalized data structure.

Text format is suitable for modest data, but it may not be a good idea to use it for big data, even though it is human-readable, because of the lack of storage optimizations and performance drawbacks. For example, the only metadata information in CSV files is in the header field but there is no type checking, so string fields could end up as numeric fields and vice versa. In JSON, there is a lot of metadata repetition taking up space and there is no native compression of repeating values.

Avro and Parquet are both binary formats, so how do we decide when to use what? Parquet (columnar) provides better query response and storage/capacity considerations, whereas Avro (row-based) has better schema evolution.

This is a good time to introduce a new open data protocol: **Delta**. A few top-level characteristics of Delta are listed here:

- It is an open source data format.
- At its core, it is Parquet plus transactional logs.
- A table built on top of the Delta data format is called a Delta table.
- A data store of Delta tables is called a Delta lake.

Delta provides reliability with rich schema validation and transactional guarantees while improving performance. We will explore Delta's capabilities in the next chapter. In the next section, we'll look at the effect of data compression on not only our storage bills but also our operational efficiencies.

# Leveraging data compression

Compression is the process of encoding information using fewer bits of data than the original representation. It is useful to save disk space and for reducing the I/O bandwidth when sending data either over the internet or from storage to RAM for processing. Compression algorithms generally come in two varieties, lossy and lossless:

- Lossy compression, as the name suggests, involves the loss of some data. It is useful for unstructured data such as images, audio, and video, where the sizes of these datasets are huge, and dropping some data is not as easily perceptible.

- Lossless, on the other hand, does not tolerate any loss of data. It is useful for the typically structured dataset, which is sensitive to even the slightest loss of data precision of the original data inserted into the database. Even a tiny change in the data stored could make it unusable. For example, financial data where a small precision change in data will manifest in a much larger number when aggregated.

**Codec** stands for **compressor/decompressor** and is used to compress and decompress digital media. Compression algorithms work in two steps. In the first step, the data is compressed in a finite amount of time. Subsequently, in the second step, while accessing the data, there is a decompression step that takes another chunk of time. The benefits of compression from cheaper storage and faster access to data should be better than the cost of performing the compression, which includes both the time to compress while storing the data and the time to decompress during retrieval of the same data. The price to pay is the amount of time that it takes to do the compression and subsequent decompression.

It is important to remember that there is a trade-off between the degree of compression and the speed of compression and subsequent decompression. Splitting a file allows for increased parallelism in the compression and decompression process. So, the splittability of a file is an important consideration as well. Formats such as `gzip` and Snappy are not splitable, whereas `bzip2` is, which also shines in terms of the degree of compression. When it comes to speed of compression, however, Snappy and LZO formats are faster than both `gzip` and `bzip2`, with `bzip2` being the slowest.

In this section, we examined different aspects of file formats and how to choose wisely to ensure faster read/write performance, reduce disk storage footprint, flexible and performant storage compression options, and support splitable files to leverage the underlying compute of distributed systems.

# Common big data design patterns

Design patterns provide a common vocabulary for data personas to understand and share design and architecture blueprints. So, given a set of requirements, everyone understands the likely design pattern to apply as a plausible solution. Traditional software engineering design patterns are object-oriented and are of three types, categorized under creational, structural, and behavioral patterns. Data engineering is not necessarily **object-oriented** (**OO**) related and is better articulated around concepts of data ingestion, transformation, storage, and analytics patterns. In the next few sections, we will look at reusable patterns in each of these areas.

# Ingestion

Ingestion refers to all aspects of consolidating data into a target site for further processing and analysis from multiple sources using different languages, different file formats, and different sizes and frequencies. The number of such combinations is large, which is why we see a lot of data vendors in this space each having its own strengths and special sauce to make this movement and transformation easy.

## Unified API

There is always a need to ingest data from multiple disparate data sources. Unifying the connector architecture to act as a seamless adapter simplifies the ingestion pattern. For example, the jdbc/odbc driver can be used to talk to any relational store. The following diagram illustrates this bridge design pattern that hides the underlying complexity and presents the user with a simple unified API:

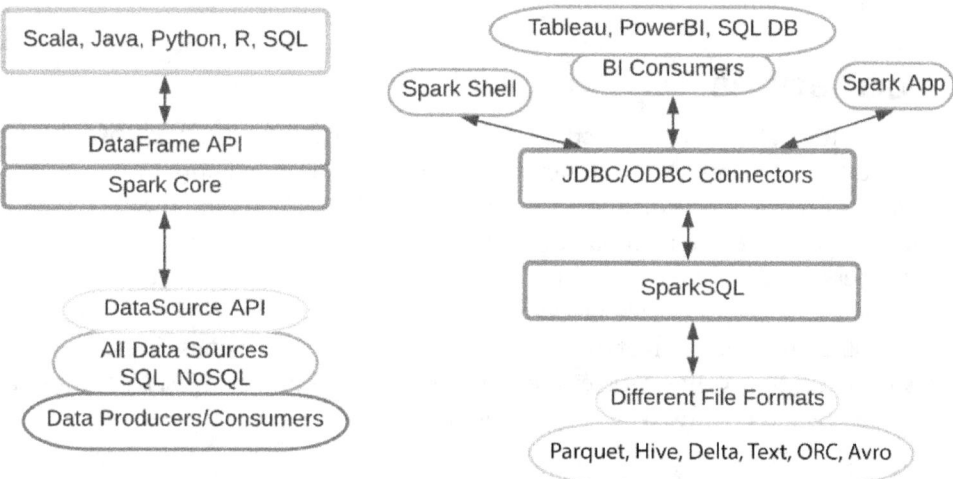

Figure 2.12 – Bridging abstractions with common APIs and the connector pattern

Spark supports multiple languages, multiple sources and sinks, multiple file formats, and programming interfaces using a single API.

## Speed layer

The Lambda architecture refers to separate batch and streaming pipelines and Kappa unifies both. There is a trend to avoid the Lambda architecture from maintaining separate pipelines in favor of the Kappa architecture, as the subsequent data reconciliation in Lambda gets difficult to maintain.

Structured streaming API constructs in Spark are very similar to batch constructs but under the covers, they do much more than batch. The idea that every batch pipeline can be expressed as a streaming pipeline controlled by a *processing trigger interval* has some inherent benefits. A lot of the heavy lifting of managing the processing state is delegated to the streaming constructs. For example, *which files have been processed and which have not* is delegated to the streaming constructs instead of having to maintain it explicitly as in batch operations.

Other benefits include *exactly once* semantics, resilience to failures, ability to tolerate some late-arriving data, in-stream analytic capabilities, and maintaining a single pipeline instead of two, which are all good reasons to consolidate pipelines. There are some additional constructs in streaming such as **checkpointing** to maintain the last offset read that helps to recover from a state of failure, **windowing** to aggregate in a specified time interval, and **watermarking** for accommodating late-arriving data within a certain threshold. It is no wonder that the Kappa architecture wins for its flexibility, ease of use, and subsequent low maintenance.

# Transformations

This refers to all the data-wrangling activities to refine and cleanse the data. This is referred to in different terms such as cleansing, munging, and remediation but in essence, it is the process of refining raw data to more usable analytic and consumption formats ready for target systems and encompasses the heavy lifting aspects of a data pipeline. The data can be diverse and complex and not necessarily always well structured. It may start out with data discovery and move on to validating, cleaning, structuring, enriching, and eventually, publishing the data. Not only do we need to write the business logic to do all this but also address the evolving system requirements of schema changes, intermittent failures, and errors and operate within the designated SLAs.

## Handling schema changes

Big data systems prefer a *schema on read*, whereas relational systems adopt *schema on write*. What this means is that a change in the schema of incoming data will be allowed in the former and gated in the latter. Although this may seem like a convenient feature, if left ungoverned, it has ramifications on the data quality, rendering it useless. So, a disciplined schema evolution with controls in place for gradual and approved changes is the preferred approach as it combines the best of both worlds.

## ACID transactions

Relational systems support **Atomicity, Consistency, Isolation, and Durability (ACID)** transactions and non-relational ones support **Basically Available, Soft State, Eventual Consistency (BASE)** properties. The inability to honor ACID has negative quality ramifications while doing **Create, Read, Update and Delete (CRUD)** operations on data at scale, which is why some big data systems, such as Delta, Hudi, and Iceberg, have extended support for it. Providing ACID transaction guarantees on big data is a preferred pattern.

## Multihop pipeline

Different use cases have different SLA requirements. A DevOps persona may consume data from the landing zone, a data scientist may consume curated data from the refined zone, and a BI analyst may consume data from an aggregated rolled-up zone. A multihop pipeline is a tiered pipeline where data is increasingly refined as it moves along the pipeline and each stage saves the curated data to disk. This pattern helps address the multiple needs of all the data personas:

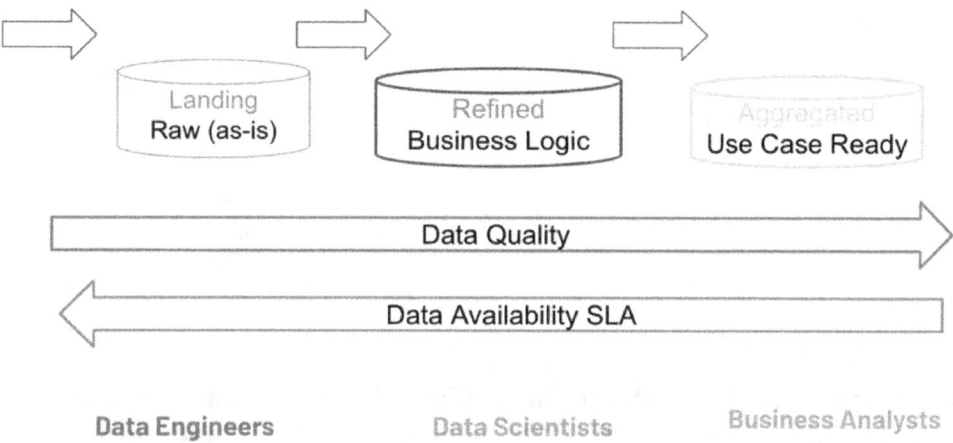

Figure 2.13 – Multihop pipeline to continuously increase data quality

There are additional advantages to using this design pattern, namely the following:

- It is more robust to failures.
- Its modular structure helps in debugging issues faster.
- It addresses different consumption needs and SLAs, as highlighted earlier.
- It facilitates collaboration so that multiple personas can work alongside each other on different parts of the pipeline.

These hops or zones are logical steps and may be referred to by different names, such as bronze, silver, and gold, but the idea behind them is the same. There may be a need to have two refined zones or maybe a use case does not need a refined zone and data moves directly from raw to aggregated.

# Persist

Persistence is the act of storing data and refers to saving data to disk in performant formats to aid in the speed of reading and writing to disk.

## Separation of compute from storage

Earlier, compute and storage were together and tightly coupled. This makes linear scaling hard. However, decoupling them gives the flexibility of switching off compute when not in use to save cost. This not only helps with system scalability but also with cost. The meter runs only when data is processed; the historical processed data may not be referenced any longer, so there is no need to keep them tied to warm compute nodes.

## Multiple destinations

There may be a need to accommodate diverse consumers with different SLA needs; the data should be forked multiple times in the pipeline.

## Denormalization

Compressed columnar storage formats and denormalized wide tables to avoid joins is the preferred pattern.

## In-stream analytics

Data coming in at a high frequency can either be landed to disk first and then consumed, or consumed in flight. The latter is, of course, more suitable for lower end-to-end latencies. Not all streaming systems support in-stream operations, but most of the sophisticated ones, including structured streaming, do. If there is an appetite to consume data for **complex event processing (CEP)**, it needs to be consumed in-stream.

## Best practices

We can summarize some of the best practices as follows:

- Understand requirements and model the data for consumption patterns.
- Profile data to understand trends and characteristics.

- Establish quality guard rails.

- Use data classification to identify sensitive data and other unique characteristics.

- Update metadata to educate other users on the dataset particulars.

- Follow established design patterns.

- Choose file format carefully.

- Compress data from the start.

- If necessary, use co-processors, for example, **field-programmable gate arrays (FPGAs)**, or specialized processors such as GPUs.

- Match compression type to data.

- Combine with data deduplication.

- Build use cases off the refined data in the curated zone.

Most businesses want access to the data as soon as possible, and since disk operations are more expensive than in-memory operations, in-stream analytics is getting increasingly popular.

# Summary

In this chapter, we talked about the importance of data modeling exercises to organize and persist the data while designing a new ETL use case so that subsequent data operations can benefit from an optimal balance of performance, cost, efficiency, and quality.

A good data model helps us with faster query speeds and reduces unnecessary I/O throughput brought about by expensive wasted scans. A design-first approach forces us to think through the data relations and can not only help reduce data redundancy but also help improve the reuse of pre-computed results, thereby reducing storage and computing costs for big data platforms. The increase in efficiency of data utilization improves the overall user experience. Having stable base datasets ensures more consistency of derived datasets further down the pipeline, thereby improving the quality of generated insights.

In the next chapter, we will look at the Delta protocol and the main features that help bring reliability, performance, and governance to the data in your data lake.

# Further reading

Please check out the following links for further reading:

- *DAS Webinar: Best Practices in Metadata Management*: `https://www.dataversity.net/das-webinar-best-practices-in-metadata-management`

- *What is metadata management?*

  `https://www.collibra.com/us/en/blog/what-is-metadata-management`

- *P-Codec: Parallel Compressed File Decompression Algorithm for Hadoop*: `https://www.researchgate.net/publication/315076898_P-Codec_Parallel_Compressed_File_Decompression_Algorithm_for_Hadoop`

# 3
# Delta – The Foundation Block for Big Data

*"Without a solid foundation, you will have trouble creating anything of value."*

*– Erica Oppenheimer, on academic mastery*

In the previous chapters, we looked at the trends in **big data** processing and how to model data. In this chapter, we will look at the need to break down data silos and consolidate all types of data in a centralized **data lake** to get holistic insights. First, we will understand the importance of the Delta protocol and the specific problems that it helps address. Data products have certain repeatable patterns and we will apply Delta in each situation to analyze the before and after scenarios. Then, we will look at the underlying file format and the components that are used to build Delta, its genesis, and the high-level features that make Delta the go-to file format for all types of big data workloads. It makes not only the data engineer's job easier, but also other data personas including ML practitioners and BI analysts benefit from the additional reliability and consistency that Delta brings. It is no wonder that it is poised to serve as a foundational block for all big data processing.

In this chapter, we will cover the following topics:

- Motivation for Delta
- Demystifying Delta
- The main features of Delta
- Life with and without Delta

# Technical requirements

The following GitHub link will help you get started with Delta: `https://github.com/delta-io/delta`. Here, you will find the Delta Lake documentation and QuickStart guide to help you set up your environment and become familiar with the necessary APIs.

To follow along this chapter, make sure you have the code and instructions as detailed in this GitHub location: `https://github.com/PacktPublishing/Simplifying-Data-Engineering-and-Analytics-with-Delta/tree/main/Chapter03`

Examples in this book cover some Databricks specific features to provide a complete view of capabilities. Newer features continue to be ported from Databricks to the Open Source Delta. (`https://github.com/delta-io/delta/issues/920`)

Let's start by examining the main challenges plaguing traditional data lakes.

# Motivation for Delta

Data lakes have been in existence for a while now, so their need is no longer questioned. What is more relevant is the specifics of the solution's implementation. Consolidating all the siloed data by itself does not constitute a data lake. However, it is a starting point. Layering in governance makes the data consumable and is a step toward a curated data lake. Big data systems provide scale out of the box but force us to make some accommodations for data quality. Age-old aspects of transactional integrity were compromised on a distributed system because it was very hard to maintain ACID compliance. Due to this, BASE properties were favored. All of this was moving the needle in the wrong direction and from pristine data lakes we were moving toward data swamps, where the data could not be trusted and hence insights that were generated on the data could not be trusted either. So, what is the point of building a data lake?

Let's consider a few common scenarios that are increasingly hard to resolve on any big data system:

- Multiple jobs read and write to the same table concurrently. Consumers build reports based on the same data and do not wish to see partial, incomplete, or bad data at any time.

- Recovering from a data pipeline failure on account of any reason, including human error, transient infrastructure glitches, insufficient resources, or changing business needs. It is easy to fix and retry the job, but it is much harder to clean up the mess that's created to a consistent state so that the retry can proceed.

- Lots of small files are produced, say, from a streaming workload, causing performance issues.

- A few big files can cause data skew and jobs to fall behind their expected schedules.

- A carefully selected and reviewed schema has been left tainted, leading to data corruption and unreadable data.

- Big data systems are designed to work with large datasets, but there is an occasional need to do fine-grained deletes and updates or merges. This needle in a haystack scenario is very prevalent in GDPR use cases, for instance.

- Recovering from a data pipeline failure, such as an error in the pipeline being detected, may require a rollback to a few days prior instantly so that new ingestions can proceed.

Different data types need to be handled differently, which is how different types of data platforms came into existence. For example, an ML platform's requirements are very different from a streaming platform's. In the next section, we will look at the primary ones and see if we can generalize towards a unified data platform.

## A case of too many is too little

Data needs over the years have evolved. Many companies/organizations start with a data warehouse, then invest in a data lake, then need a machine learning platform, and then use a separate system to handle streaming needs. Very soon, this complexity gets out of hand, and stitching these fragile systems increases the end-to-end latency, which means that the quality of the data suffers as it is moved around multiple times. Here, the company/organization needs to be on a single platform that can handle a majority of these data needs and must resort to a specialized system in rare cases when it is truly warranted.

Being on a next-generation data platform means that all your data personas can be accommodated with their choice of tools and preferences – these personas include data engineers, data scientists, data analysts, and DevOps personnel. They typically work in one of the following three areas at various levels of sophistication:

- BI reporting and dashboarding
- Exploratory data analysis and interactive querying
- Machine learning at various levels of sophistication

The following diagram shows an example of a utopian data platform that accommodates the needs of all data personas and every use case that they work on:

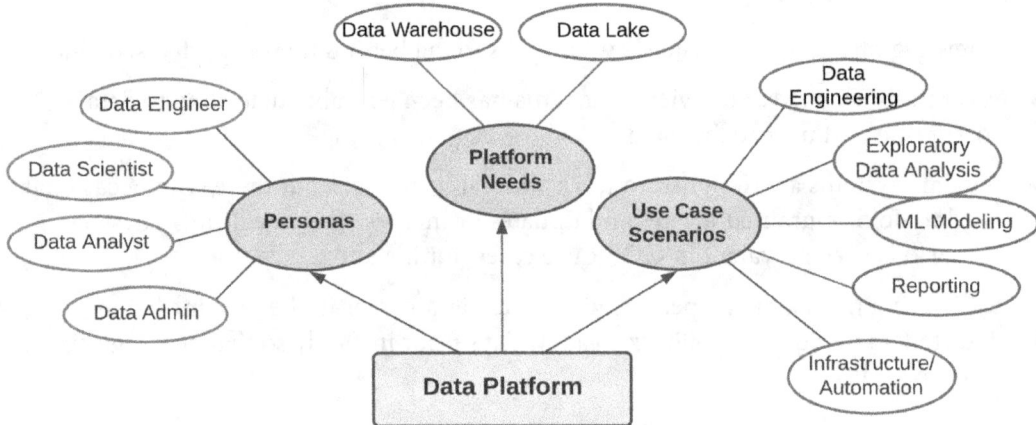

Figure 3.1 – Unified data platform

Platforms are shells on top of data formats and data query engines. The better the data format, the more performant the engine that's built on top will be and the more robust the platform housing it. So, it suffices to say that the underlying data protocol and format is an important consideration. Wouldn't it be nice to have a single data format in the open protocol that can support an organization's needs for batch, streaming, reporting, and machine learning needs?

# Data silos to data swamps

Data lakes help avoid the need for **data silos**. However, inadequate governance can turn them into **data swamps**. The following diagram is a perfect analogy of the swings where we may start with a data silo. If we swing too hard, we may find ourselves in a data swamp, which is just as bad, if not worse:

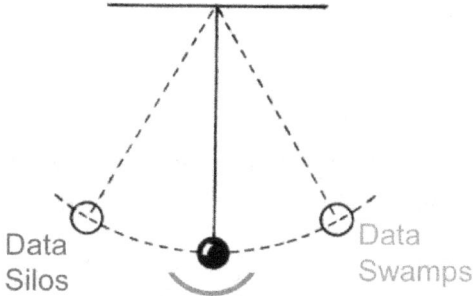

Figure 3.2 – Data silos can easily turn into data swamps

A **data silo** is an isolated source of data that is only accessible to a single **line of business** (**LOB**) or department. It leads to inefficiencies, wasted resources, and obstacles in the form of incomplete data profiles and the inability to construct deep insights. Let's examine the reasons why data silos are created in the first place as it is important to understand what to avoid:

- **Structural**: An app built for a specific purpose where data sharing is not a core requirement.

- **Political**: A sense of proprietorship over a system's data so that it is not readily shared with others, either on account of the data being sensitive or the team being possessive.

- **Growth**: Newer technology that is incompatible with existing systems and datasets, leading to siloed systems.

- **Vendor Lock-In**: Technology vendors gate data access to customers, so once you've invested in the technology of that vendor, it is very difficult to wiggle free and the technical baggage carries over.

On the other hand, a **data swamp** is a large body of data that is ungoverned and unreliable. It is hard to find data and even harder to use it, which is why it's often used out of context. This is the opposite of data silos in the sense that the data is there and has been brought together, but because it has been done without adequate process and policy, it is as good as not being there. That would be a wasted investment. So, how can we ensure our data silos evolve into data lakes and instead of ending up as data swamps, blossom into data reservoirs?

# Characteristics of curated data lakes

Curated data lakes build upon four pillars that have a direct impact on insight generation. They are as follows:

- **Reliability**: Data consistency accelerates the pace of innovation as data personas have more time to focus on the use case instead of grappling with data reconciliation and quality issues.

- **Performance**: A performant system translates directly into a lower **total cost of ownership** (**TCO**) with a simple architecture that is easy to develop and maintain.

- **Governance**: Access to data based on privileges is necessary for regulatory compliance because a single security breach can cause permanent long-term credibility damage.

- **Quality**: Improving data quality improves the quality of insights.

In the previous chapter, we explored various file formats and concluded that binary formats with good compression and flexible schema handling are ideal candidates for big data systems. Parquet is an established format and Apache Spark operations are optimized on it. However, there are several inadequacies for which complex solutions have been built on top of Parquet that are both difficult to develop and maintain. Let's look at some of these inadequacies in the context of the following concepts:

- DDL and DML operations
- Schema evolution
- Unifying batch and streaming workloads
- Time travel
- Performance

We will take up the challenges that arise in each of these areas when we discuss the main features of Delta. But first, let's look at its DDL and DML capabilities and understand the underlying format to appreciate the thought process that inspired its creation.

# DDL commands

The **data definition language** (**DDL**) includes commands that allow you to create and drop a database or table. Let's take a look.

## CREATE

With Parquet, you must run the REPAIR TABLE command to force statistics to be computed before the ingested data can be made visible via SELECT queries. This is not needed for Delta. The following example shows this action when using a Parquet table:

1. Load data into a Parquet table from a given path:

```
spark.sql("""
    CREATE TABLE IF NOT EXISTS some_parquet_table
    USING parquet
    OPTIONS (path = '{}')
""".format(dataPath))
```

2. Read the data right after the ingestion shows 0 records in the table:

```
SELECT COUNT(*) FROM some _ parquet _ table
```

Let's understand why the count shows 0 records right after ingestion. There are two paths to be aware of – one is the data path that's identified by dataPath and the other is the metastore path for some_parquet_table. This has to do with a strategy called **schema on read**, which was popularized by Hadoop and other NoSQL databases to make the ingestion process super flexible, allowing any schema to be written. Here, the onus of sorting the mess was pushed downstream to the consumers. This is in sharp contrast to most of the traditional relational database technologies, where the schemas were strictly enforced and followed the **schema on write** paradigm. The schema is checked when the data is written, so if there is a discrepancy, it throws an error and it is detected right at the source during the ingestion process.

**MetaStore ChecK (MSCK)** helps update the metastore datapaths with the counts so that subsequent queries will see the right number of records. The following steps show how to run the MSCK command:

1. Run the REPAIR TABLE command to update the metastore data:

```
MSCK REPAIR TABLE some _ parquet _ table
```

2. The following SELECT query will now show the correct record count:

```
SELECT COUNT(*) FROM some _ parquet _ table
```

There is no need to run the REPAIR TABLE command when you're working with the Delta format. So, here, you get the benefit of *schema on read*, which is additional schema flexibility, thus allowing the schema to evolve and grow over time. It helps support the ELT path, where you keep the original data as-is and load the data before fully comprehending all the fields or are forced to drop fields to comply with predefined schemas. This is especially important for semi and unstructured data types. So, performing the preceding steps in Delta looks as follows:

```
spark.sql("""
    CREATE TABLE IF NOT EXISTS some_delta_table
    USING delta
    OPTIONS (path = '{}')
  """.format(dataPath))

# Alternate way of using DataFrameWriter api
df.write.format("delta").saveAsTable("some_delta_table")

# This select query shows the correct record count right from the get go
spark.sql("SELECT COUNT(*) FROM some_delta_table")

#  Another alternate way to make the same query is to use the file path
# This is handy when you do not want to create an external table explicitly
spark.sql("SELECT COUNT(*) FROM delta.`<dataPath>`")
```

This helps you avoid issues when developers forget to run MSCK and see incorrect data counts, which means you spend less time debugging. Delta takes that pain away by giving you correct counts right away and eliminates the need for you to use administrative commands such as MSCK to repair the table. In the next section, we will examine Delta in the context of DML commands.

# DML commands

**DML** stands for **data manipulation language**. Operations such as UPDATE, DELETE, and others qualify as DML statements.

# APPEND

The same repair table problem is encountered when appending data. The counts of the new records will require similar treatment. This applies to partition names as well since they are also a part of the metadata. Let's see how seamless this would be with Delta:

```
# Load some CSV data into a deltaPath from a given inputPath
# The data is immediately available for consumption
# There is no need to run any repair table commands
(spark.read
  .csv(inputPath)
  .write
  .mode("append")
  .format("delta")
  .partitionBy("Field1")
  .save(deltaPath) )

# Record Count displays correctly
spark.sql("SELECT count(*) FROM delta.`{}`").format(deltaPath)

# Partition information displays correctly
spark.sql("SHOW PARTITIONS some_delta_table")
```

Using append operations is a strong pattern in big data on account of new information constantly being added. Although updates are less common, they are also an important pattern and need to be handled efficiently to satisfy common CRUD operations when existing data needs to be updated, corrected, or deleted altogether.

# UPDATE

Update operations are crucial to CRUD operations when you need to perform basic data manipulation commands but are not supported in Parquet format. It is supported on Delta. The following code snippet demonstrates an error scenario where non-Delta tables are being used:

```
UPDATE parquet _ table SET field1 = 'val1'  WHERE key = 'key1'
```

This will throw an error that states `Error in SQL statement: AnalysisException: UPDATE destination only supports Delta sources.`

Now, let's try the same query but this time on a Delta table:

```
UPDATE delta_table SET field1 = 'val1'  WHERE key = 'key1'
```

Similar to the previous query, this query is honored and the number of rows that have been affected is returned as `num_affected_rows`.

## DELETE

Like `UPDATE`, delete operations are not supported in Parquet but they are supported in Delta. This simple operation has several applications and it is especially critical for GDPR compliance, where as an organization, you are expected to respond swiftly to a customer's request to delete all their personal information across all tables over the entire time range that you have been collecting information in the data lake. The following code snippet demonstrates the error scenario when non-Delta tables are used:

```
DELETE FROM  parquet_table  WHERE key = 'key1'
```

This preceding code will throw an error that states `Error in SQL statement: AssertionError: assertion failed: No plan for DeleteFromTable`. The same operation, when performed on a Delta table, will be honored:

```
DELETE FROM  delta_table WHERE key = 'key1'
```

Here, the same query that we used previously is honored and the number of rows affected is returned as `num_affected_rows`.

# MERGE

More complex data operations include handling MERGE commands as a single step as opposed to multiple steps. A merge operation is sometimes referred to as an upsert, which is a combination of an update and an insert in a single operation. This is especially useful for handling **slowly changing dimensions (SCD)** operations where the dimension table data is either overwritten or maintained with history. Merging involves a target table and an incoming source table where the two are joined on a match key. Depending on the match result, records can be added, updated, or deleted in the target table. There can be any number of WHEN MATCHED and WHEN NOT MATCHED clauses, as shown in the following snippet:

```sql
%sql
MERGE INTO target_table_identifier [AS target_alias]
USING source_table_identifier [<time_travel_version>] [AS source_alias]
ON <merge_condition>
[ WHEN MATCHED [ AND <condition> ] THEN <matched_action> ]
[ WHEN MATCHED [ AND <condition> ] THEN <matched_action> ]
[ WHEN NOT MATCHED [ AND <condition> ]  THEN <not_matched_action> ]
%sql
MERGE INTO target_deta_table A
USING new_incoming_delta_table B
ON A.uniqueId = B.uniqueId
WHEN MATCHED
  THEN UPDATE SET *
WHEN NOT MATCHED
  THEN INSERT *
```

In this section, we looked at all the data operations that would not only benefit from using Delta but also be greatly simplified. For example, there is no need to remember to run MSCK commands, nor worry about an occasional fat finger operation that corrupts data in one pipeline and destroys the pristine quality of the target data lake that several stakeholders read from. Delta allows us to maintain a single pipeline, so there is no need to maintain separate batch and streaming pipelines, nor worry about schemas growing out of control. In the next section, we will look under the covers to see what components go into creating the Delta protocol and give it its superpowers.

# Demystifying Delta

The Delta protocol is based on Parquet and has several components. Let's look at its composition. The transaction log is the secret sauce that supports key features such as ACID compliance, schema evolution, and time travel and unlocks the power of Delta. It is an ordered record of every change that's been made to the table by users and can be regarded as the single source of truth. The following diagram shows the sub-components that are broadly regarded as part of a Delta table:

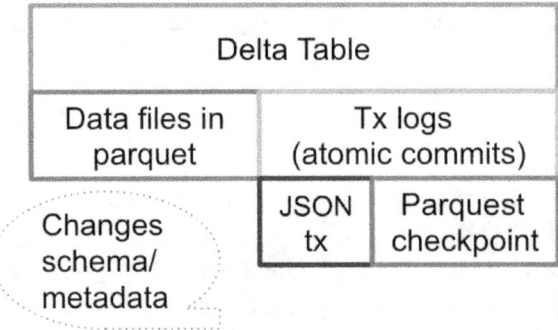

Figure 3.3 – Delta protocol components

The main point to highlight is that metadata lies alongside data in the transaction logs. Before this, all the metadata was in the metastore. However, when the data is changing frequently, it would be too much information to store in a metastore, and storing just the last state means lineage and history will be lost. In the context of big data, the transaction history and metadata changes are also big data by themselves. A metastore still holds some basic metadata details but most of the meaty details are in the transaction log.

## Format layout on disk

Both the data and the metadata are persisted on disk. Let's look at their formats more closely:

- Data remains in open Parquet format. This binary columnar format is excellent for analytic workloads:

  - If data is partitioned, each partition appears as a folder.

- Transaction details are in a `_delta_log` directory and provide ACID transactions:

  - Every commit is captured in JSON files. **Cyclic redundancy check (CRC)** files hold key data statistics and help Spark optimize queries.

  - Checkpoint files are in `.checkpoint.parquet` format and are generated after every 10 commits.

The following diagram shows what a table that's been partitioned on the `partition1` column would look like if it happened to be a Delta table:

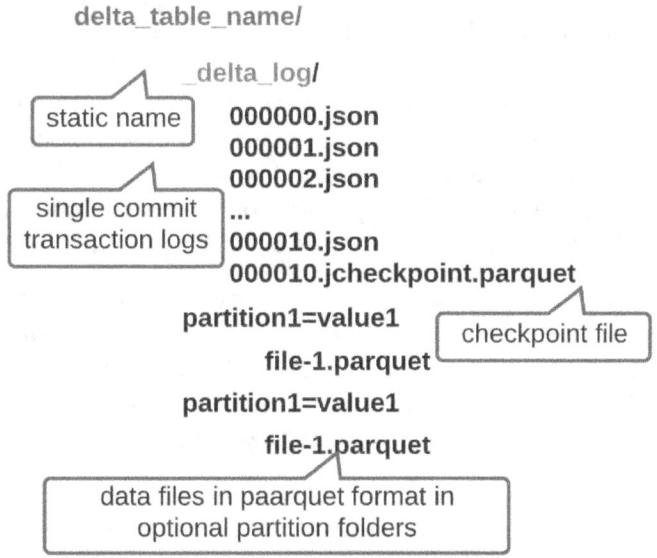

Figure 3.4 – Delta file layout on disk

If you are examining the files on disk, one quick way to check if a data path or table is Delta or not is to look for the presence of a `_delta_log` folder underneath the main data path. A non-Delta table does not have this folder.

In this section, we examined the components that comprise Delta. In the next section, we will look at the main reusable macro features that aid the creation of robust data pipelines. These features can be viewed as design blueprints. Only if the underlying format and protocol support them can they be leveraged as-is. Other systems that do not support Delta will have to write additional code to ensure these capabilities are created; in Delta, they come out of the box.

# The main features of Delta

The features we will define in this section are equivalent to weapons in an arsenal that Delta provides so that you can create data products and services. These will help ensure that your pipelines are built around sound principles of reliability and performance to maximize the effectiveness of the use cases built on top of these pipelines. Without any more preamble, let's dive right in.

## ACID transaction support

In a cloud ecosystem, even the most robust and well-tested pipelines can fail on account of temporary glitches, reinforcing the fact that a chain is as strong as its weakest link and it doesn't matter that a long-running job failed in the first few minutes or the last few minutes. Cleaning up the subsequent mess in a distributed system would be an arduous task. Worse still is the fact that partial data has now been exposed to consumers who may use it in their dashboards or models to arrive at wrong insights and trigger incorrect alarms. Thankfully, Delta, with its ACID properties, comes to the rescue. This capability refers to the fact that either the entire job succeeds or it fails, leaving no debris behind.

Delta does this using Reader Writer isolation capabilities. This means that while an ingest is happening, the partial data is not yet visible to any consumer. So, if that workload fails, no damage is done and the job can be retried. This also has other benefits. For example, multiple writers could be running workloads writing to the same table independently. Here, a hiccup in one of them wouldn't corrupt all the data in that table, nor would it affect the other workloads.

There is no special construct specification. This capability comes out of the box and helps ensure data quality so that the insights that are generated on top of them are sound as well.

## Schema evolution

In the previous chapter, we emphasized the need to undergo a data modeling exercise to fully understand the data before rushing to implement your use case on top of it. As part of this exercise, the underlying schema is established and data pipelines are constructed around it. These well-defined and well-curated schemas will evolve. Some fields will get added, others modified, and yet others dropped. We do want to accept and adapt to some of these changes, especially early on in the pipeline, and we may want to have a stricter lockdown further down the pipeline. Both Schema Enforcement and Schema Evolution are important requirements. This requires having different knobs to turn at different points in the pipeline. In the early landing stages, you want to be more flexible (Schema Evolution) and in the curated zones, you want the schema to be honored (Schema Enforcement) to prevent it from breaking your data contracts with your consumers.

This is where Delta's `mergeSchema` property comes in handy. By default, it is off, so making changes to the existing schema will throw an error. You can turn it on so that it's more flexible:

```
option("mergeSchema", "true")
or
spark.databricks.delta.schema.autoMerge.enabled
```

It allows you to make compatible changes, such as redefining an `INT` field as a `STRING` field, upcasting a field, adding new fields, and more. However, it does not allow incompatible changes to be made, such as turning a text field into a float.

Another check that helps with data governance is the concept of enforcing constraints. This automatically verifies the quality and integrity of new data that's been added to a Delta table:

- The `NOT NULL` constraint ensures that those specific columns always contain data.

- The `CHECK` constraint ensures that the Boolean expression evaluates to `true` for each row:

  ```
  ALTER TABLE <table> ADD CONSTRAINT validDate CHECK (dt >
  '2022-01-01');
  ```

Running `DESCRIBE DETAIL` or `SHOW TBLPROPERTIES` on the Delta table will list all the constraints. The `DROP CONSTRAINT` command can be used to drop an existing constraint.

Databricks Delta has built an edge feature called **Delta Live Tables**, where you can declaratively build a pipeline. The `expect` operator is used to expose the expectations on these constraints so that actions can be taken, such as dropping data or failing the pipeline.

## Unifying batch and streaming workloads

The Lambda architecture advocates managing two streams of workloads – one batch and the other streaming. This has its fair share of challenges concerning data reconciliation. The Kappa architecture challenges the need to maintain two pipelines and instead proposes maintaining a single streaming pipeline where batch is a special case of streaming. Delta adopts this stance and unifies both batch and streaming pipelines. Some people may argue that their need is batch, so why venture into streaming?

Note that business owners have an increasing appetite for real-time insights and the number of streaming workloads continues to rise. The tunable processing trigger dial allows you to easily transition, for example, from a 24-hour interval to a 5-minute interval. This way, the workload is designed only once and maintained as a single set instead of two. There are some bookkeeping activities that structured streaming constructs automatically take care of that can benefit a batch operation. For example, keeping track of the processed files using a checkpoint location takes the onus from the data engineer to the platform.

# Time travel

There are times when we need to create a snapshot of a dataset to run a repeatable workload and this can either be a data engineering or a machine learning workload. We end up making copies of copies and then forgetting what the source was, which leads to unnecessary storage and wasted cycles recreating the lineage. At times, a processing change or human error may occur that causes bad data to enter production tables. Here, we need a quick way to identify the affected records and roll them back so that data can be reprocessed correctly. In a big data system, it is very hard to do this.

Thankfully, Delta provides a time travel capability out of the box to do just this. Every batch or micro-batch operation that's performed on the data is automatically given a version and a timestamp. So, instead of maintaining multiple copies of the data in different tables, you can use the same table and provide a version number or timestamp to reference the relevant dataset. So, if multiple people are building models to compare, they all have the same view of the data and hence the model comparison is consistent.

The following code demonstrates using the time travel capabilities of Delta using the TIMESTAMP AS OF, VERSION AS OF constructs in both SQL and the DataFrame API commands:

```
#using timestamp
spark.sql("SELECT * FROM my_deta_table TIMESTAMP AS OF '2021-10-18T22:15:12.013Z'")

#using a version number
spark.sql("SELECT * FROM delta.`/delta/my_deta_table` VERSION AS OF 123")

#using a DataFrameReader API
df2 = spark.read.format("delta").option("versionAsOf", 123).load("/delta/my_deta_table")
```

All of these are additional value features that Delta brings to the table out of the box. However, if these are not executed in a performant manner, then the value will not be fully realized in a big data ecosystem. In the next section, we will look at the additional performance benefits that accompany Delta.

# Performance

All the data reliability features we mentioned previously will not be as useful if access to them is not performant. Delta enjoys all the advantages of Parquet and several more features such as data skipping, Z-Order, and delta cache. Unlike RDBMS systems, there are no secondary indices in big data systems, which primarily rely on partitions to reduce large data scans. However, a table cannot have too many partitions; at most, it can have around three because each partition works as a sub-folder. This works well for date and categorical fields, but it is not well suited for high cardinality columns such as ZIP codes.

## Data skipping

The **data skipping** feature comes out of the box with Delta and is orthogonal to partition pruning and can work alongside it. It can be applied to query scenarios of the **column op literal** format. It is sometimes referred to as I/O pruning and works by maintaining column-level statistics. All the data operations that are applied to the table use these statistics during query planning time to skip files and avoid wasted I/O.

## Z-Order clustering

Currently, this is a Databricks Delta feature. Z-Order involves using a clustering approach and can be layered on top of data skipping to make it even more effective. Data in Delta tables can be physically altered using a built-in command called OPTIMIZE. It can be applied to the entire table or certain portions, as specified in a WHERE clause. For example, if there is a rolling duration of 2 months that users query, there is no need to optimize the entire table. This is effective when the workload consists of equally frequently related single-column predicates on different columns. The Z-Order syntax is as follows:

```
OPTIMIZE <table> [WHERE <partition_filter>]
ZORDER BY (<column>[, …])
```

## Delta cache

Currently, this is a Databricks Delta feature. In Spark, caching is done at the RDD or DataFrame level by calling cache() or persist().

Delta offers file-based caching for Parquet files. This means that a remote file, once fetched, can be cached on the worker nodes (provided it supports NVMe/SSD) for faster access. The results of the query are cached as well. The following three configurations are handy in this regard:

- `spark.databricks.io.cache.maxDiskUsage`: This specifies the reserved disk space in bytes per node for cached data.

- `spark.databricks.io.cache.maxMetaDataCache`: This specifies the reserved disk space in bytes per node for cached metadata.

- `spark.databricks.io.cache.compression.enabled`: This is used if the cached data is stored in a compressed format.

Caching can be turned on and off and can be controlled explicitly for `SELECT` queries:

```
CACHE SELECT column_name[, column_name, ...] FROM [db_name.]
table_name [ WHERE boolean_expression ]
```

Now that we have established the benefits of Delta, let's look at how easy it is to convert your existing workloads into Delta.

# Life with and without Delta

The tech landscape is changing rapidly, with the whole industry innovating faster today than ever before. A complex system is hard to change and is not agile enough to take advantage of the pace of innovation, especially in the open source world. Delta is an open source protocol that facilitates flexible analytic platforms as it comes prepackaged with a lot of features that benefit all kinds of data personas. With its support for ACID transactions and full compatibility with Apache Spark APIs, it is a no-brainer to adopt it for all your data use cases. This helps simplify the architecture both during development as well as during subsequent maintenance phases. Features such as the unification of batch and streaming, schema inference, and evolution take the burden off DevOps and data engineer personnel, allowing them to focus on the core use cases to keep the business competitive.

It is very easy to create a Delta table, store data in Delta format, or convert it from Parquet into Delta in place. For non-Parquet formats such as CSV or ORC, a two-step process needs to take place. This consists of converting the data into Parquet format and then into Delta, as follows:

```
%sql
CONVERT TO DELTA [ [db_name.]table_name | parquet.`<path-to-table>` ]
[NO STATISTICS]
[PARTITIONED BY (col_name1 col_type1, col_name2 col_type2, ...)]
```

The following table demonstrates the ease of creating a Delta table by comparing it with the standard way of using the Parquet format:

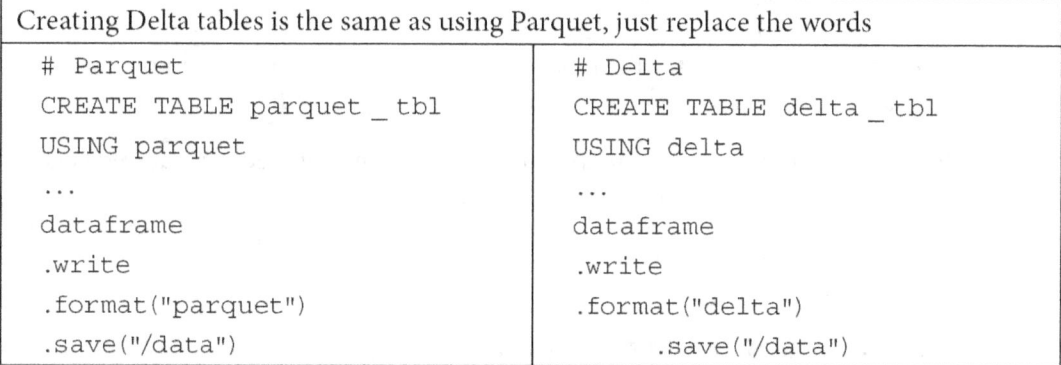

| Creating Delta tables is the same as using Parquet, just replace the words | |
|---|---|
| ```# Parquet\nCREATE TABLE parquet_tbl\nUSING parquet\n...\ndataframe\n.write\n.format("parquet")\n.save("/data")``` | ```# Delta\nCREATE TABLE delta_tbl\nUSING delta\n...\ndataframe\n.write\n.format("delta")\n.save("/data")``` |

Table 3.1 – Creating a Delta table compared to using Parquet

If you have existing Parquet tables, they can be converted in place into Delta format, as shown here:

| Using Delta with existing Parquet tables |
|---|
| ```# Step 1: Convert Parquet to Delta\nCONVERT TO DELTA <parquet path or table>\n[PARTITIONED BY (<partition columns if any>)]``` |
| ```# Step 2: Optimize layout for fast queries: in addition to above conversion\nOPTIMIZE <delta_table>\nWHERE date >= current_timestamp() - INTERVAL 1 day\nZORDER BY (<z order column/s>)``` |

Table 3.2 – Converting existing Parquet tables into Delta tables

At the beginning of this chapter, we introduced a few problematic scenarios. Let's revisit them to see how Delta addresses each one:

- Multiple jobs read and write to the same table concurrently. Consumers build reports on the same data and do not wish to see partial, incomplete, or bad data at any time:

  - Delta provides Reader Writer isolation, which means that until a write is completely done, no consumer sees any of the partial data and if one of the ingestion pipelines fails, it does not affect the other pipelines. This means that multiple pipelines can write to a single table concurrently and not step on each other.

- Recovering from a data pipeline failure on account of any reason, including human error, infrastructure transient glitches, insufficient resources, or changing business needs. It is easy to fix and retry the job, but it is much harder to clean up the mess that's created to a consistent state so that the retry can proceed:

  - Delta's ACID transactions ensure that either everything succeeds or nothing gets through. This ensures there is no mess to clean up afterward.

- Lots of small files produced, say, from a streaming workload, causing performance issues:

  - Operations such as OPTIMIZE help with file compaction.

- A few big files can cause data skew and jobs to fall behind their expected schedules:

  - Splitable file formats and repartitioning strategies help with large file issues.

- A carefully selected and reviewed schema left tainted, leading to data corruption and unreadable data:

  - Disciplined schema evolution controls help determine when to allow or disallow changes to the underlying schema, allowing only the benign and necessary merges while controlling the undesirable ones from corrupting consumption layer tables.

- Big data systems are designed to work with large datasets, but there is an occasional need to do fine-grained deletes and updates or merges. This needle in a haystack scenario is very prevalent in GDPR use cases, for instance:

  - Updates, deletes, and merges are provided as atomic operations in the Delta format, allowing for painless modifications to be made to data at scale.

- Recovering from a data pipeline failure, such as an error in the pipeline being detected, may require a rollback to a few days prior instantly so that new ingestions can proceed:

  - Time travel allows you to easily roll back the dataset using either a timestamp or a version number to control where to rewind to.

In the next section, we will look at the role of Delta as a foundational piece for the Lakehouse architecture, which promotes diverse use cases on top of a single view of data in an open format, allowing multiple tools of the ecosystem to consume it.

## Lakehouse

Lakehouse is a new architecture and data storage paradigm that combines the characteristics of both data warehouses and data lakes to create a unified basis for all types of use cases to be built on top of it. There is no need to move data around. Data is curated and remains in an open format and serves as the **single source of truth** (**SSOT**) for all the consumption layers. A modern data platform has needs that span traditional data warehouses, data lakes, machine learning systems, and streaming systems and there is some overlap among these systems. A Lakehouse offers features that span all four systems, as shown in the following diagram:

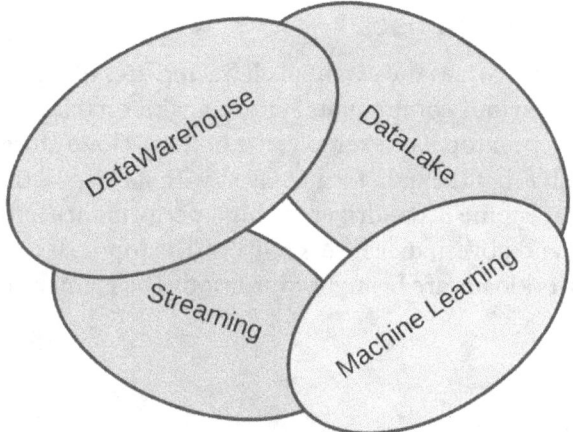

Figure 3.5 – The common use cases in a Lakehouse borrow the leading characteristics from disparate data systems

It imbibes the structure and governance that's inherent to data warehouses, the flexibility and cost-effectiveness of data lakes, the low latency and high resiliency of streaming systems, and the ability to run different ML tools and frameworks. Sometimes, the streaming and ML components are considered to be part of the data lake, so it is usually termed as the best of breed of data lakes and data warehouses.

The following diagram captures the high-level characteristics of a Lakehouse architecture and shows how the data stack is built up:

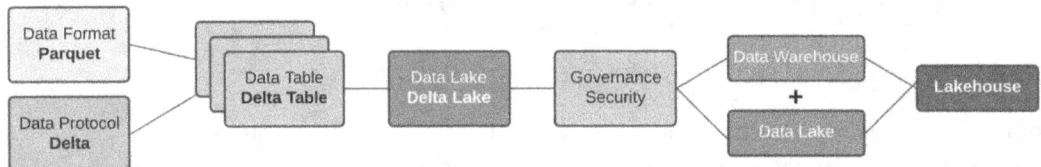

Figure 3.6 – Delta is the foundational block for a Lakehouse

A data lake can help break a data silo, but by itself, it is not a Lakehouse. Delta is the most important ingredient as it forms the foundational layer that allows each of these characteristics. Together with security and governance, the full potential of the architecture is unleashed.

## Characteristics of a Lakehouse

The following diagram summarizes the essential characteristics of a Lakehouse architecture and how the various components stack up, with each layer adding a critical feature to support the layer on top. For example, the bottom-cloud storage layer offers scalability and affordability that the data format layer, with its transactional capabilities, is layered on. The execution engine is the driver for high performance and feature execution and is what the security and governance layers rely on. The topmost layer is where the applications of diverse workloads are brought to fruition. The primary characteristics of a Lakehouse are as follows:

- Transaction support
- Schema enforcement and governance
- BI support
- Storage is decoupled from compute
- Openness
- Support for diverse data types ranging from unstructured to structured data
- Support for diverse workloads

- End-to-end streaming:

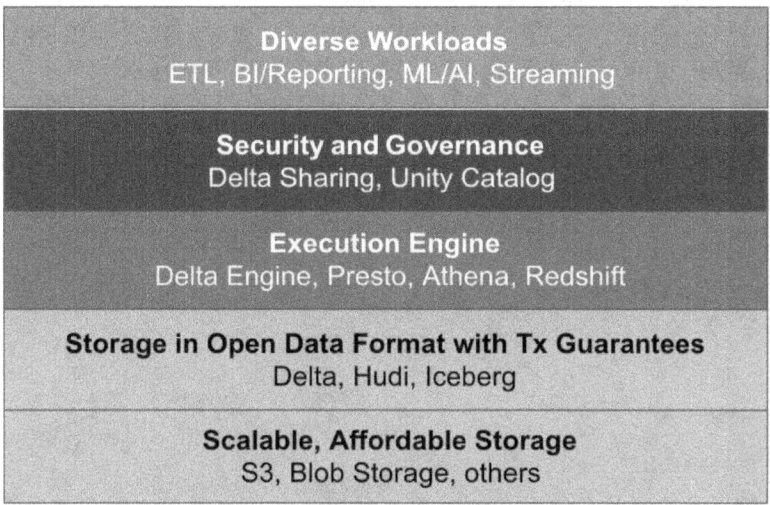

Figure 3.7 – Lakehouse example

The Databricks platform is an example of a Lakehouse architecture – the bottom-most layer represents cloud storage, where data resides in Parquet format; right beside it is the Delta protocol; the Delta Execution engine brings in the optimizations, governance is achieved via the Unity catalog; secure data sharing is achieved via Delta sharing; and the diverse workloads are built right on top, covering all the aspects of data engineering, data science, BI, and AI.

By eliminating the need to maintain both a data warehouse and a data lake, there are efficiencies in resource utilization and a direct cost reduction. Some of the immediate benefits of a Lakehouse architecture are the unification of BI and AI use cases. BI tools can now connect directly to all the data in the data lake by pointing the query engine at it. There is reduced data and governance redundancy by eliminating the operational overhead of managing data governance on multiple tools.

However, there are a few notes of caution:

- What should we do with existing data warehouses? In *Chapter 7, Delta for Data Warehouse Use Cases*, we will look at this in detail as we compare and contrast the use cases that are built on these different data platforms.

- Are the tools mature and unified? Hundreds of enterprises are using this as part of their production architecture, thereby proving it is ready for prime time.

- Are we moving toward a monolith again? Not really – this is a paradigm and can be used as a blueprint for different lines of business to create a data mesh of lakes or hub and spoke models with the central hub offering core capabilities and the spoke refining their definitions, thereby reusing administrative and governance definitions of the hub with the flexibility to refine their business-specific implementations.

The following diagram captures the challenges of traditional data lakes in four categories – that is, reliability, performance, governance, and quality:

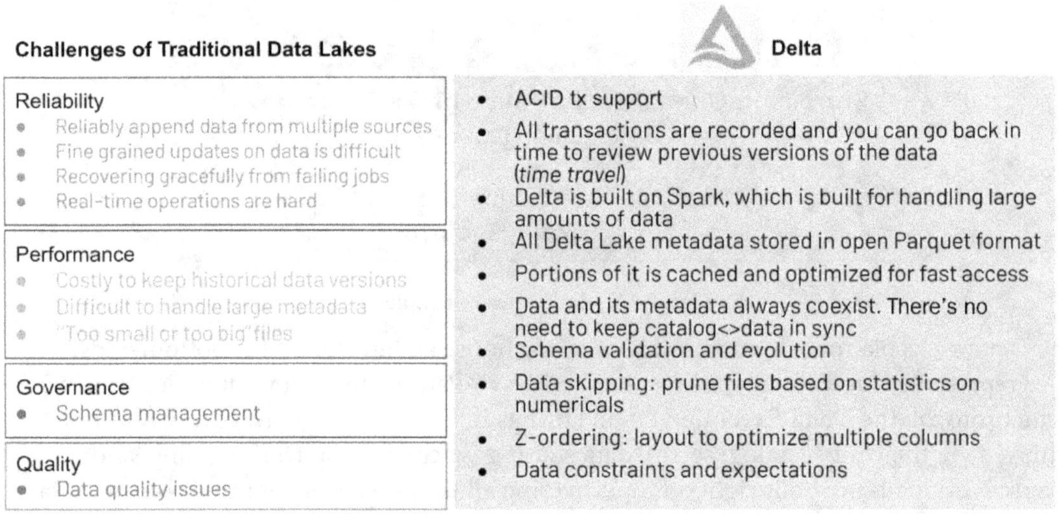

**Challenges of Traditional Data Lakes**   **Delta**

**Reliability**
- Reliably append data from multiple sources
- Fine grained updates on data is difficult
- Recovering gracefully from failing jobs
- Real-time operations are hard

**Performance**
- Costly to keep historical data versions
- Difficult to handle large metadata
- "Too small or too big" files

**Governance**
- Schema management

**Quality**
- Data quality issues

- ACID tx support
- All transactions are recorded and you can go back in time to review previous versions of the data (*time travel*)
- Delta is built on Spark, which is built for handling large amounts of data
- All Delta Lake metadata stored in open Parquet format
- Portions of it is cached and optimized for fast access
- Data and its metadata always coexist. There's no need to keep catalog<>data in sync
- Schema validation and evolution
- Data skipping: prune files based on statistics on numericals
- Z-ordering: layout to optimize multiple columns
- Data constraints and expectations

Figure 3.8 – Challenges of traditional data lakes that Delta addresses

The right-hand side of the preceding diagram lists the characteristics of the Delta protocol that help address these gaps and pave the path toward simplification and innovation, all while keeping the openness of the format and API to avoid getting into a vendor lock-in trap. In the following chapters, we will examine each in more detail.

# Summary

Delta helps address the inherent challenges of traditional data lakes and is the foundational piece of the Lakehouse paradigm, which makes it a clear choice in big data projects.

In this chapter, we examined the Delta protocol, its main features, contrasted the before and after scenarios, and concluded that not only do the features work out of the box but it is very easy to transition to Delta and start reaping the benefits instead of spending time, resources, and effort solving infrastructure problems over and over again.

There is great value when applying Delta to real-world big data use cases, especially those involving fine-grained updates and deletes as in the GDPR scenario, enforcing schema evolution, or going back in time using its time travel capabilities.

In the next chapter, we will look at examples of ETL pipelines involving both batch and streaming to see how Delta helps unify them to simplify not only creating but maintaining them.

# Section 2 – End-to-End Process of Building Delta Pipelines

In this section, you will learn how Delta aids the plumbing and heavy lifting of raw data to refine it for analytic purposes. Instead of spending time, effort, and resources on additional code to address everyday data challenges, data staff can now focus primarily on the business use case. Delta provides constructs for greater reliability, quality, and performance of your data and is becoming the de-facto standard for big data projects as it greatly simplifies the tasks of all the data staff involved in creating a data product or service.

This part includes the following chapters:

# 4
# Unifying Batch and Streaming with Delta

*"We are only as strong as we are united, as weak as we are divided".*

*– J.K. Rowling, author of the Harry Potter series*

In the last chapter, we examined Delta's capabilities and how it solves the challenges of traditional data lakes to give you curated data that is the foundation for sound insights without having to solve common operational problems over and over again. In this chapter, we will look at the two patterns of ingestion in data systems, namely, **batch** and **streaming**. Traditionally, they would have required two separate pipelines and the associated cost and effort to create, maintain, and reconcile data between the two pipelines. Thanks to protocols such as Delta, these two pipelines can now be consolidated.

In particular, we will be covering the following topics:

- Moving toward real-time systems
- Streaming ETL
- Handling streaming scenarios
- Trade-offs in designing streaming architectures
- Streaming best practices

As enterprise workloads move toward real-time ingestion and insight generation, understanding nuances around streaming architectures is imperative for a data persona. Unlike a batch process, which is run at predefined intervals, giving you a safety net to resort to during non-operational time slots, streaming is continuous and adjustments to pipelines and the graceful handling of error scenarios all have to be handled in parallel, which makes it very challenging. Systems running 24*7 have cost implications that need to be considered carefully so that the value of getting early insights is balanced with the additional cost of infrastructure.

In this chapter, we will go over the streaming concepts in the context of Delta and how unifying both batch and streaming using a Kappa architecture is the preferred approach to constructing modern data and ML pipelines.

# Technical requirements

To follow along this chapter, make sure you have the code and instructions as detailed in this GitHub location: `https://github.com/PacktPublishing/Simplifying-Data-Engineering-and-Analytics-with-Delta/tree/main/Chapter04`

Examples in this book cover some Databricks specific features to provide a complete view of capabilities. Newer features continue to be ported from Databricks to the Open Source Delta.

Let's get started!

# Moving toward real-time systems

As their names suggest, **batch** is a form of periodic ingestion of data, whereas **streaming** is a process where data ingestion is either continuous or in micro-batches. There is no denying that the trend is toward the real-time ingestion, analysis, and consumption of data. This gives rise to the question of why is every pipeline not a streaming one?

There may be several producers of data for the same target table. Some may be fast-moving, while others could be slower. If the nature of your data is such that it comes once a month, then we certainly do not want to have compute running more frequently than once a month from a cost savings perspective. Hence, some folks may say that cases such as these force us to have batch ingestion. In this chapter, we will present an argument to justify that batch is actually a type of streaming workload and that all workloads can be expressed as a streaming pipeline. You may argue, 'Isn't streaming more complex?' Counterintuitive as it may sound, it actually isn't. The initial adoption and ramp-up may take some time, but the streaming constructs do a lot more under the covers, especially around processing state management that benefits the batch. So yes, we will still maintain the stance that it is more pragmatic to think of all ingestion as streaming.

The Spark API constructs for batch and streaming are very similar and structured streaming APIs support both, as shown in the following snippets. Delta is the underlying storage that unifies the two as well:

1.  Read data once from a given location:

    ```
    val inputDF = spark.read.json(<input cloud storage path>)
    ```

2.  Perform operations using the standard DataFrame API:

    ```
    inputDF.groupBy($"action", window($"time", "1 hour")).count()
        .writeStream.format("delta")
        .save(<output cloud storage path>)
    ```

3.  Read data continuously from the same location:

    ```
    val inputDF = spark.readStream.json(<input cloud storage
    path>")
    ```

4.  Perform operations using the standard DataFrame API:

    ```
    inputDF.groupBy($"action", window($"time", "1 hour")).count()
        .writeStream.format("delta")
        .start(<output cloud storage path>)
    ```

Structured streaming treats the incoming data sources as an endless stream that can be visualized as an unbounded table on which queries can be executed to compute results tables that can be persisted to one or more sinks in one of three modes: **append**, **complete** (complete replace, as in a kill and fill scenario), and **update**. Spark uses a process of incrementalization to convert the batch-like query to a streaming execution plan. A trigger is a control that determines at what frequency the pipeline wakes up, as shown in the following diagram. Spark maintains the state for all the micro-batches of data that flow through.

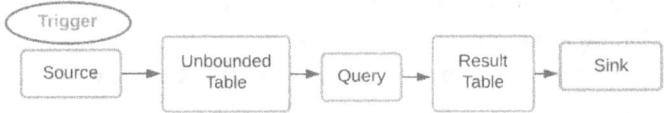

Figure 4.1 – Unifying batch and stream processing in structured streaming

The following diagram shows an **Extract Transform Load** (ETL) architecture blueprint. Data flows in from various data sources. The ingestion block encapsulates unified batch and streaming processing. The heavy lifting is done to wrangle the data by applying various business transformations. This is where, using a medallion architecture, Delta tables refine data in bronze, silver, and gold zones to get them into the hands of the AI/BI personas who wield different tools to consume it as appropriate to their use cases.

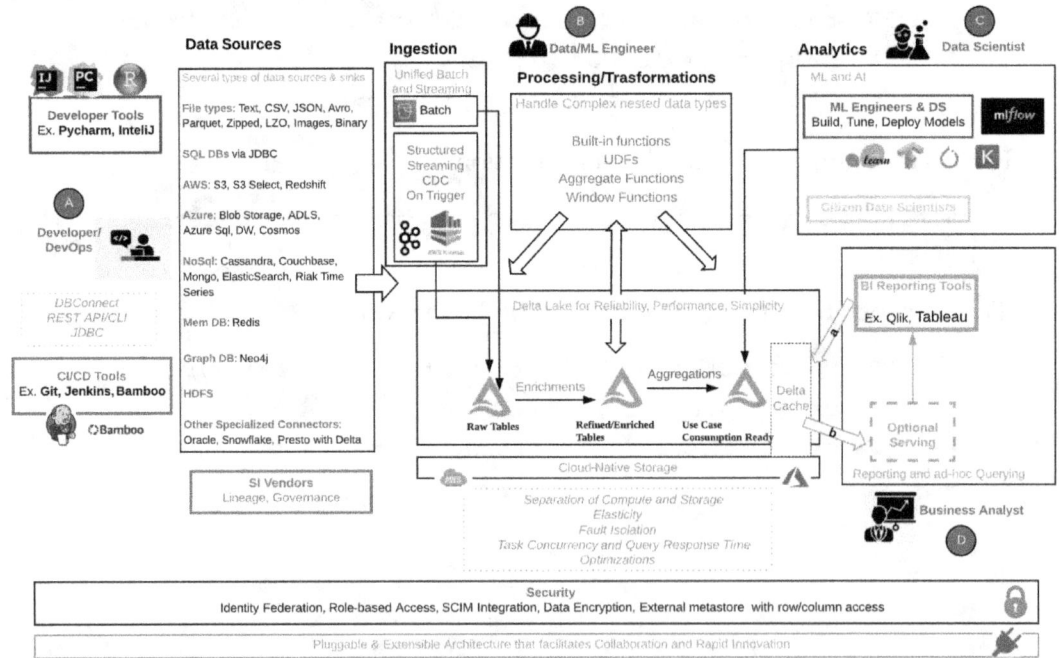

Figure 4.2 – ETL architecture blueprint

Structured streaming is integrated into Spark's Dataset and DataFrame APIs; in most cases, you only need to add a few method calls to run a streaming computation.

## Streaming concepts

Before we go into details of streaming architectures and stream processing, it is important to understand some of the common concepts around streaming, such as checkpointing, watermarking, processing trigger, and others, as shown in the following table:

| | |
|---|---|
| Stream Components | Every streaming pipeline has three elements, namely:<br>1) A source<br>2) A sink, also called a target<br>3) A processing engine |
| File-based versus Event based | **File-Based**: Landing data to disk first and then ingesting<br>**Event-Based**: Processing on the fly in-stream without first landing on disk |
| Micro-batch versus continuous streaming | **Micro-Batch:** Buffer data for a few sec or ms<br>**Continuous:** As soon as data is received |
| Processing trigger | The frequency or interval at which data is crunched |
| Output modes | Three options exist in writeStream to save data to disk, namely<br>1) **append**: data is always added<br>2) **complete**: data is completely replaced, aka, "kill and fill"<br>3) **update**: the relevant records are updated |
| Checkpoint | A mechanism to recover from failure by remembering the last known data offset processed |
| Window | Aggregation interval for counting, averaging, and more for ready-to-consume metrics |
| Watermark | The period of time during which the late arrival of data can be tolerated before closing the gates and dropping data that arrives after the defined tolerance, that is, not processing late-arriving data |
| Stream Operations | All transformations to wrangle streaming data |
| In-Stream Analytics | All transformations on the moving data while data is in flight |

Figure 4.3 – Streaming concepts

In the next section, we will look at the two popular ETL data processing architectures and justify which one is the better choice.

## Lambda versus Kappa architectures

We had covered Lambda and Kappa architectures in the previous chapter in the context of big data design patterns. Lambda architecture promotes having two separate pipelines to handle batch and streaming. Lambda is the older of the two and was a product of its time as streaming architectures were not as evolved and batch was the preferred mode The serving layer taps into data from views computed from both layers. The Kappa architecture promotes unification of the batch and speed layers. The industry trend is to adopt Kappa and that is exactly what Delta makes provision for out of the box. Jay Kreps popularized Kappa in 2013 during his LinkedIn years when monolithic relational systems were being broken down into multiple distributed systems and the architectures were all log-focused (reference: `https://engineering.linkedin.com/distributed-systems/log-what-every-software-engineer-should-know-about-real-time-datas-unifying`).

Logs are time-ordered journals that capture what happened to the data and are very powerful as they offer a wide range of benefits, ranging from aiding the implementation of **ACID (Atomicity Consistency Isolation Durability)** transactions to supporting data migrations, to creating deterministic and consistent outcomes. Kreps states that *tables* support *data at rest*, while *logs* capture *change*. In the context of a bank account, you may have a current balance, but using the past transaction history in the ledger you can reconstruct the path of credits and debits done to get there. Another way to visualize it would be through the familiar Git repository, where every user's code changes are versioned, helping you to track and roll back to a specific version if the need arises. The following diagrams highlight the difference:

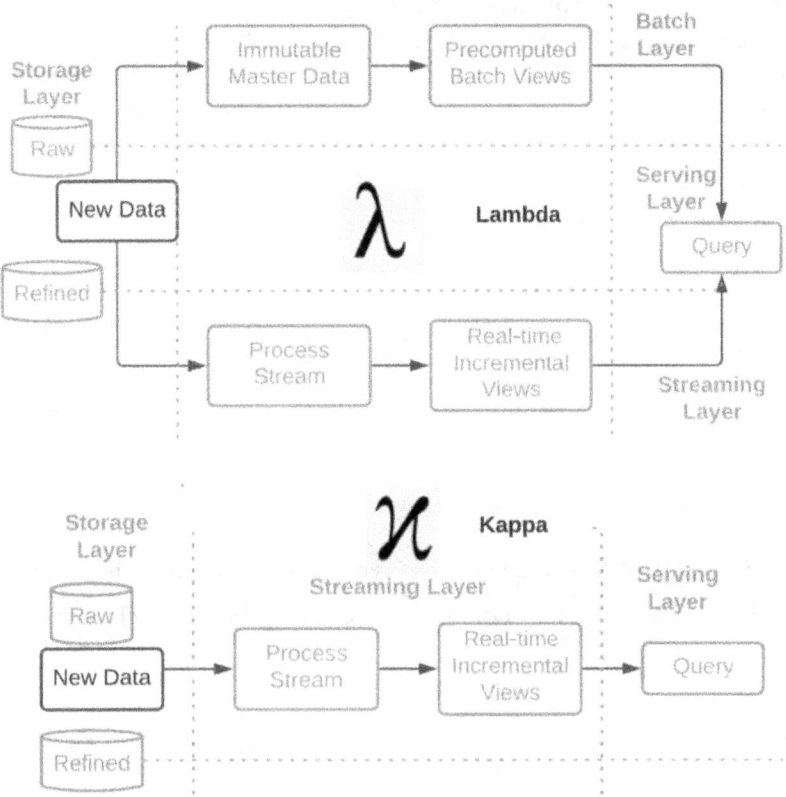

Figure 4.4 – Lambda versus Kappa architectures

The concept of Reader/Writer isolation is necessary in an environment with multiple producers and consumers involving the same data assets. The idea is for multiple pipelines to work independently with the promise that bad or partial data is never presented to the user. The scenario here is that of a single target table fed by multiple batch and streaming jobs pulling data from a host of data producers. Any one of these pipelines can fail at any time. That should not cripple the other pipelines. If a pipeline fails, the others are insulated from it and can continue to operate.

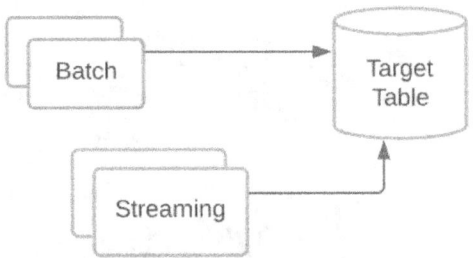

Figure 4.5 – Reader/Writer isolation is a key requirement of multi-tenant processing

The following table captures the main differences between the two architectures and Delta is based on a Kappa architecture:

| Lambda | Kappa |
|---|---|
| Two separate processing layers – batch and streaming | Batch is treated as a form of streaming<br>It is simpler and doesn't need a separate batch layer |
| Duplication of data processing | Single processing pipeline to develop and maintain<br>Horizontally scalable; fewer resources, less reconciliation efforts required, and, hence, less downtime |
| Batch is reliable whereas streaming is approximate | Single pipeline that is reliable<br>Streaming with consistency |
| Works better for historical load re-processing | Works well with a high-speed stream processing engine to enable end-to-end low latency |

Figure 4.6 – Lambda versus Kappa

In the next section, we will look at the different types of streaming.

# Streaming ETL

Streaming use cases comprise three main categories of real-time applications – decision engines and alerting apps; BI analytics and tools, such as SQL and search engines; and data science and ML use cases, as highlighted in the following diagram:

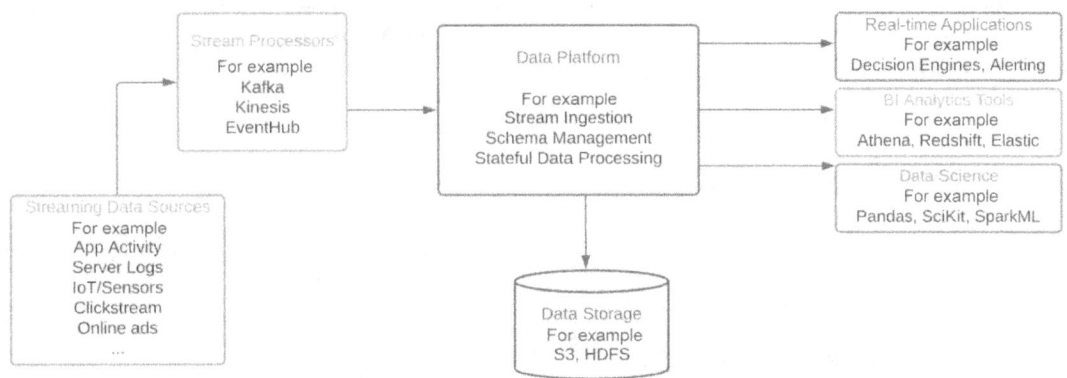

Figure 4.7 – Streaming use cases

In the next section, we will look at the three stages of **ETL (Extract, Transform, Load)** as it relates to streaming.

## Extract – file-based versus event-based streaming

There are two types of stream processing – **file-based** and **event-based**. The former applies to data that has landed on disk, and the latter to data in flight, and which typically requires a streaming service such as Kafka, Kinesis, or EventHub from which `spark.readStream` consumes the data. For example, a Kafka cluster consists of several brokers monitored by Zookeeper. Data is stored in topics that are broken down into one or more partitions that allow for scalability, fault tolerance, and higher parallelism. A stream partition is an ordered list of data records that, in turn, constitute key-value pairs. Needless to say that the latter has faster end-to-end latency at the cost of more moving parts, as shown in the following diagram:

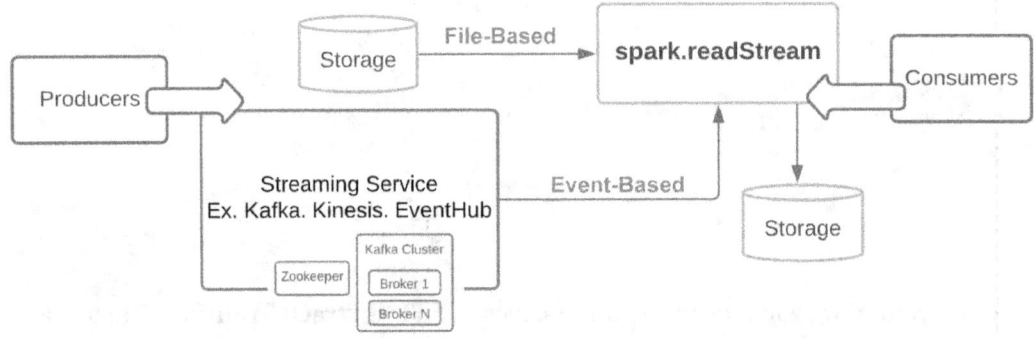

Figure 4.8 – File-based versus event-based stream processing

This section talked about the initial ingest process where stream data enters the system either via files or events. In the next section, we will look at the various transformations that are done to refine the data.

# Transforming – stream processing

Event-based processing used message brokers in the first generation. This included **message-oriented middleware** (**MOM**) such as ActiveMQ, RabbitMQ, and the like. The next-generation processing to support very high performance with persistence moved on to stream processors including Apache Kafka and Amazon Kinesis. These typically have a massive capacity of several gigabytes per second and are tightly focused on streaming with little support for data transformations or task scheduling. The following diagram shows an architecture leveraging MOM:

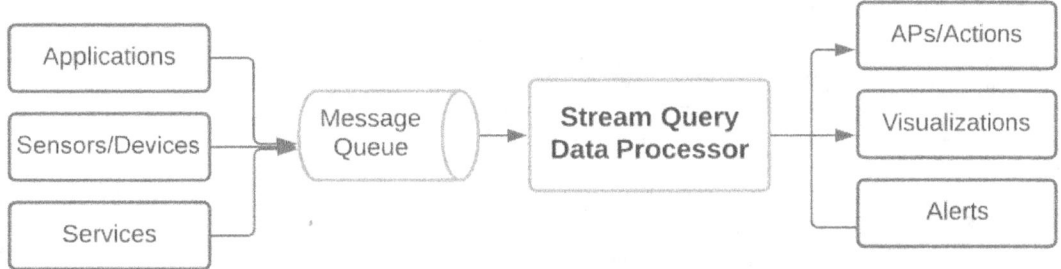

Figure 4.9 – MOM-based event processing

The following is a more modern adaptation of the previous architecture using Kafka, which is a popular streaming service:

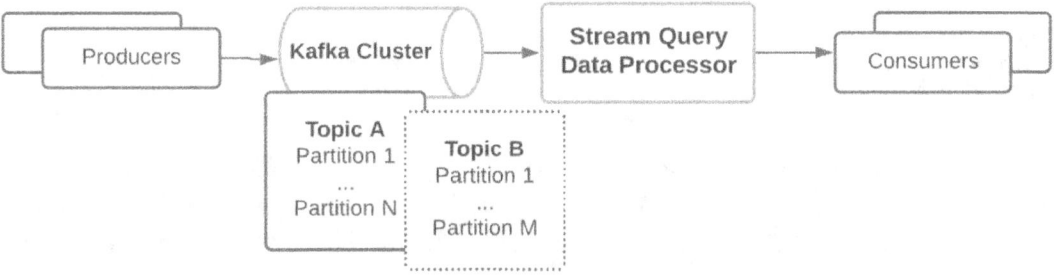

Figure 4.10 – Streaming services-based event processing

Structured streaming offers APIs to do all the common transformations on the data, including filtering, aggregating, joining, and many more in-stream as the data is in flight. Here are two examples:

- An example of reading from Kafka:

```python
# Example of reading from Kafka
from pyspark.sql.functions import col
kafkaServer = "<host:port>"                              # Specify Host & Port & the name of the topic
topicName = 'iot-topic'

iotData = (spark.readStream                              # Get the DataStreamReader
  .format("kafka")                                       # Specify the source format as "kafka"
  .option("kafka.bootstrap.servers", kafkaServer)        # Configure the Kafka server name and port
  .option("subscribe", topicName)                        # Subscribe to the Kafka topic
  .option("startingOffsets", "latest")                   # stream to latest when we restart notebook
  .option("maxOffsetsPerTrigger", 1000)                  # Throttle Kafka's processing of the streams
  .load()
  .repartition(8)
  .select(col("value").cast("STRING"))
)
```

- An example of reading from Kinesis to pick up IoT device data reporting on a user's fitness metrics:

```python
# Example of reading from Kinesis to pick up iot device data reporting on a user's fitness metrics
from pyspark.sql.functions import *
kinesisDF = spark.readStream \
  .format("kinesis") \
  .option("streamName", "kinesis-stream") \
  .option("region", "us-east-2") \
  .option("initialPosition", "trim_horizon") \
  .load()

dataDF = kinesisDF.select(col("data").cast('string').alias("data"))
dataDF.createOrReplaceTempView("stream_data")

stream_df = spark.sql("""
    SELECT data:id,
    data:user_id,
    data:device_id,
    cast(data:num_steps as int) as num_steps,
    cast(data:miles_walked as double) as miles_walked,
    cast(data:calories_burnt as double) as calories_burnt,
    cast(data:timestamp as timestamp) as timestamp
    FROM stream_data""")

(stream_df.writeStream.format("delta")
    .trigger(processingTime='30 seconds')
    .option("checkpointLocation", <checkpoint location in storage>)
    .outputMode("append")
    .table("device_data_streaming"))
```

Delta can be both a sink and a source for streams where data arrives continuously, in frequent batches. In the next section, we will look at persisting strategies.

## Loading – persisting the stream

In addition to in-stream transformations that can be chained, it is necessary to eventually persist the data to either a relational data store, a message broker, or a Data Lake. Here are some trade-offs among these options:

| Streaming Data Storage | Advantages | Disadvantages |
| --- | --- | --- |
| Data Warehouse or traditional relational data store | Easy to use SQL operations | Scalability challenges, proprietary formats and could get expensive. |
| Leverage storage of the Message Broker | No additional setup and can accommodate any schema without explicitly having to define the schema. | The data retention period is limited, can be several-fold more expensive than traditional data storage, and is a good solution for recency use cases |
| Data Lake cloud storage | Has all the same advantages as message brokers with increased scalability and reduced cost. | Higher latency for some use cases |

Figure 4.11 – Pros and cons of different streaming storage strategies

It is important to iterate that data lake storage is the clear winner, with its affordable, scalable, and open format capabilities. The others can supplement in edge cases. In the next section, we will examine several scenarios typical of streaming use cases and how to address each of them.

# Handling streaming scenarios

In this section, we will see how to tackle common streaming requirements such as joining a stream with other data, some of which could be a mix of batch and streams; recovering from an intermittent failure scenario, which may involve restarting the stream after a period of inactivity; and handling late-arriving data, among other scenarios.

# Joining with other static and dynamic datasets

Joins are a very common operation and streaming datasets are no exception. Very often, they need to be joined with other datasets, usually a slowly-changing dimension dataset to fortify the data for rich analytics. Let's consider an IoT use case where devices are being manufactured in lots and getting registered in a Delta lookup table. As the devices are deployed in the field, they start emitting sensor data that is streaming in nature and comes at a higher velocity. This IoT data needs to be joined with the device's lookup data. Storing device data in Delta allows for data versioning and change detection, which enables data to be reloaded/refreshed without the need for a restart. Had it been a non-Delta table, then an update table will not automatically do so. The batch job can just as easily be changed to streaming, meaning that the dimension table is updated whenever new data shows up without the need to explicitly schedule it.

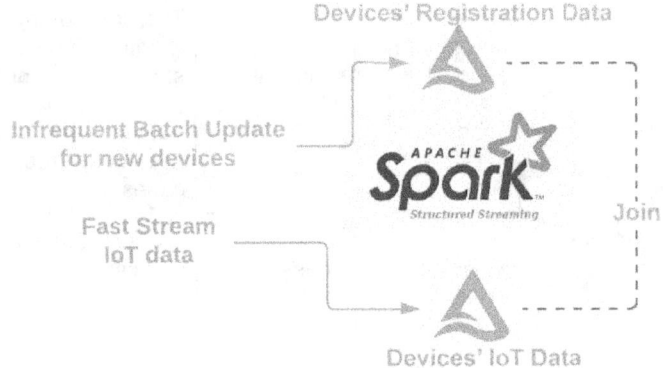

Figure 4.12 – Stream-to-stream joins

The following code snippet demonstrates how to join two streams without having to save either to disk first:

```
#hold the lookup table in a dataframe
devices_df = spark.table("devices_lookup_tbl")

#read the incoming iot data
iot_df = spark.readStream() ….

#join the 2 dataframes on device identifier
join_df = iot_df.join(devices_df, ['device_id])

#persist to disk
join_df.writeStream \
  .format('delta') \
  .outputMode('append') \
  .option('checkpointLocation', checkpoint_path) \
  .toTable(""devices_iot_tbl")

#At this point if new devices get registered, the devices_df will get the updates
#Pre-delta, this would be stale data and would require the user to re-read the table each time prior to a join
```

In the next section, we will look into strategies for recovering from failures.

# Recovering from failures

Delta leverages the semantics and constructs of Spark Structured Streaming. The format ('delta') signifies that the underlying protocol is delta. Spark Structured Streaming is a scalable, fault-tolerant streaming engine built on top of the Spark SQL engine that ensures 'exactly once' semantics, meaning that in the event of failure, data is neither dropped nor duplicated. It does this by using the concepts of checkpointing and 'write ahead logs'. To enable checkpointing, set the `checkpointLocation` option to a cloud storage path before you start the query, as shown in the following example. Upon restarting the query after a failure, it will continue where the failed one left off, guaranteeing fault tolerance:

```
#ETL in Spark Structured Streaming

spark.readStream                                    # Extract
.option("sep", ";")
.csv(<input data path>)

.selectExpr("cast(age as int) age")                 # Transformation
.select(col("income").cast('double').alias("income"))
.writeStream                                        # Load
.format("delta")
.option("path", <output data path>)

.trigger("5 minutes")                               # Controls the frequency of processing
.option("checkpointLocation", <checkpoint path>)  # tracks progress of the query
.start()
```

The preceding code snippet demonstrates a periodic stream ingestion pipeline, a subsequent transformation, and a subsequent load to disk in Delta format.

# Handling late-arriving data

Structured streaming uses a windowing strategy to handle key streaming aggregations: windows over event-time and late and out-of-order data. Structured streaming supports three kinds of windowing operations, namely, **tumbling**, **sliding**, and **session windows**. Tumbling is a fixed size and non-overlapping, sliding is similar, but with the overlap, and session has a dynamic size. A session window remains open if subsequent events are received and will close following a designated period of inactivity. **Watermarking** is a concept used to limit state while handling late data. For example, you expect data in a 5-minute interval, and if it is late by 2 minutes, it is still acceptable, but not more than 2 minutes. The following diagram illustrates the differences:

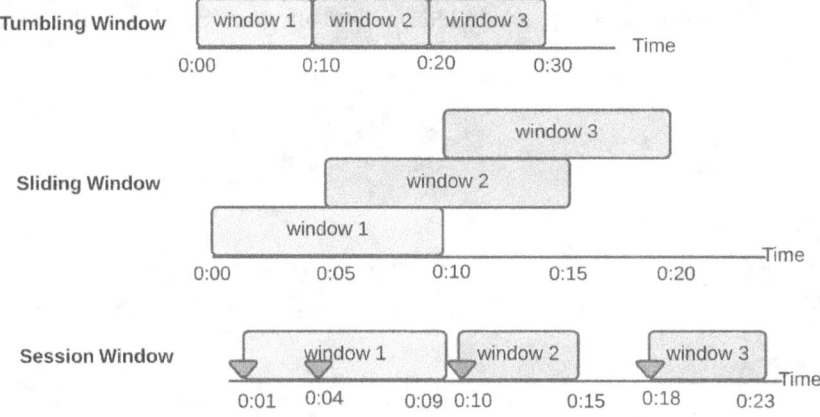

Figure 4.13 – Windowing operations in structured streaming

The following snippets demonstrate the differences in the three approaches:

```
# tumbling window
windowedCountsDF = \
  eventsDF \
    .withWatermark("eventTime", "10 minutes") \
    .groupBy("deviceId", window("eventTime", "10 minutes") \
    .count()

# sliding window
windowedCountsDF = \
  eventsDF \
    .withWatermark("eventTime", "10 minutes") \
    .groupBy("deviceId", window("eventTime", "10 minutes", "5 minutes")) \
    .count()

# session window
windowedCountsDF = \
  eventsDF \
    .withWatermark("eventTime", "10 minutes") \
    .groupBy("deviceId", session_window("eventTime", "5 minutes")) \
    .count()
```

For example, count over 5-minute tumbling (non-overlapping) windows on the eventTime column in the event as in the following example. Spark SQL will automatically keep track of the maximum observed value of the eventTime column, update the watermark, and clear the old state:

```
from pyspark.sql.functions import *

#Tumbling window
windowedAvgSignalDF = \
  eventsDF \
    .withWatermark("eventTime", "10 minutes") \
    .groupBy(window("eventTime", "5 minute")) \
    .count()

#Overlapping window specified by window length and sliding interval.
windowedAvgSignalDF = \
  eventsDF \
    .withWatermark("eventTime", "10 minutes") \
    .groupBy(window("eventTime", "10 minutes", "5 minutes")) \
    .count()
```

Streaming is analogous to an unbounded table, which brings up the question of maintaining state. Some operations are stateless, but others require state to be managed.

## Stateless and stateful in-stream operations

Stateful operations include aggregations, stream-stream joins, dropping duplicates, and others where the data received previously matters. There are use cases for **Complex Event Processing (CEP)** or to maintain user sessions over a definite or indefinite period of time and persist those sessions for post-analysis. Using mapGroupsWithState and flatMapGroupsWithState APIs (java and scala), you can implement customized stateful aggregations beyond event-time basics and event-time processing.

Below is the definition of a mapping function that maintains an integer state for string keys and returns a string. Additionally, it sets a timeout to remove the state if it has not received data for an hour.

```
// A mapping function that maintains an integer state for string keys and returns a string.
// Additionally, it sets a timeout to remove the state if it has not received data for an hour.

def mappingFunction(key: String, value: Iterator[Int], state: GroupState[Int]): String = {

  if (state.hasTimedOut) {                 // If called when timing out, remove the state
    state.remove()

  } else if (state.exists) {               // If state exists, use it for processing
    val existingState = state.get          // Get the existing state
    val shouldRemove = ...                  // Decide whether to remove the state
    if (shouldRemove) {
      state.remove()                        // Remove the state

    } else {
      val newState = ...
      state.update(newState)               // Set the new state
      state.setTimeoutDuration("1 hour")   // Set the timeout
    }

  } else {
    val initialState = ...
    state.update(initialState)             // Set the initial state
    state.setTimeoutDuration("1 hour")     // Set the timeout
  }
  ...
  // return something
}

dataset
  .groupByKey(...)
  .mapGroupsWithState(GroupStateTimeout.ProcessingTimeTimeout)(mappingFunction)
```

In this section, we went over stream transformations and typical patterns to handle such as duplicates, and late-arriving data, among others. In the next section, we will understand how to handle design trade-offs iteratively to provide a solution that not only addresses functional and non-functional requirements, but at a cost that is palatable to a client.

# Trade-offs in designing streaming architectures

Spark has multiple ways of achieving the end goal with tunable performance, cost, and quality. Hence, there is a need for a process to understand the goals/requirements of use cases. For each goal, define the strategy needed; for each goal, define the resources required; compute the cost of resources employed. The process is repeated until expectations are balanced. This is where trade-offs need to be considered, as shown in the following diagram; either the goal or resource has to be tweaked:

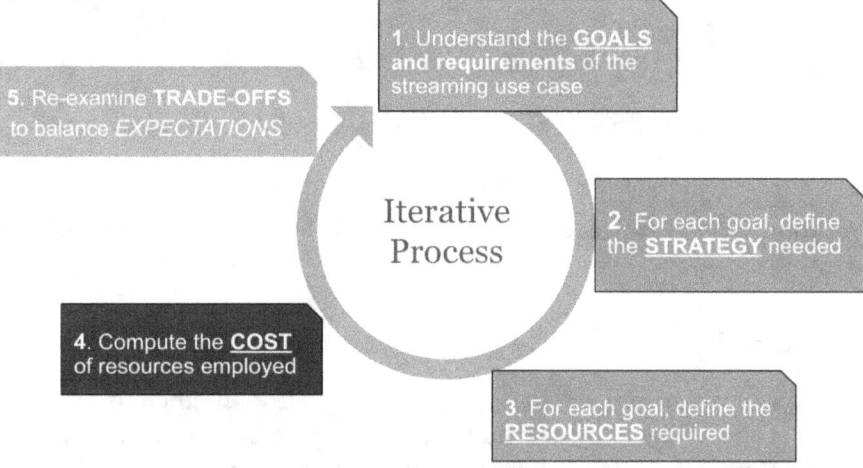

Figure 4.14 – Balancing streaming service requirements and resources

It is a balancing act of managing the various goals with the resources that the team is willing to bring to the table. Goals refer to requirements regarding scalability, performance, processing logic, quality, reliability, and availability. Resources refer to compute, storage, and integration services, and the effort needed to not only create but also maintain the solution. An imbalance here will lead to unmet requirements or an expensive solution and, in both cases, the use case runs the risk of being shelved.

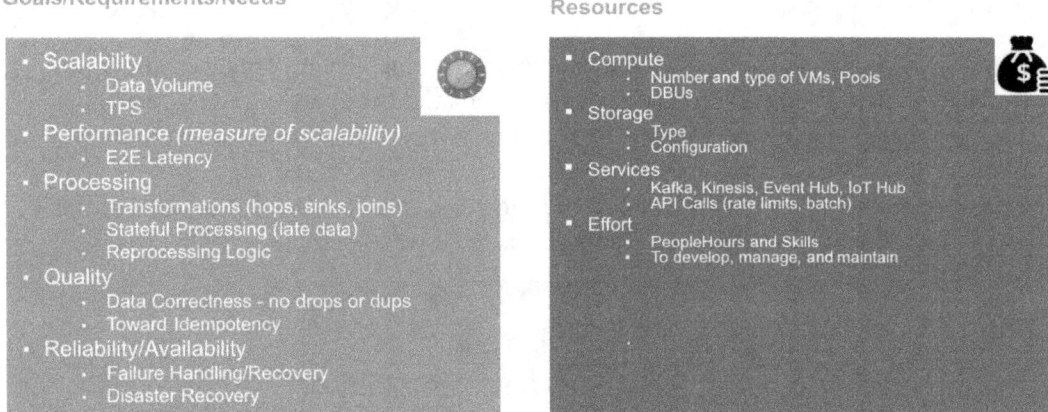

Figure 4.15 – Requirements versus resources

In the next section, we will look at examples in terms of demonstrating the trade-off considerations.

# Cost trade-offs

Let's consider a network threat detection use case. The goal/requirement is to ingest data from EventHub at a low frequency, and transform and aggregate the data. Models are applied to predict threat detection patterns and suggest recommendations. The primary customer pain was identified as "After spending way too much on storage yesterday, we shut down 24/7 processing this morning. Storage was 70%, with virtual machines accounting for 30%". In addition, it was noted that the egress costs were unusually high, while low-streaming volume resulted in lots of very small files, resulting in high compute cost, while storage access costs were high as well.

It is strange that storage costs would exceed compute costs! These are some things to look out for:

- **Storage**: When data needs to be archived, 'cool' storage is used since it is less expensive than 'warm'. Cool storage refers to storage that is rarely accessed, whereas warm data is naturally in active use and gets accessed more frequently.

- **Access**: However, if data needs to be accessed frequently, access from 'cool' storage is more expensive. Active data should be retained in 'warm' storage.

- **Data replication**: Local replication is cheaper than global replication. So, if disaster recovery can tolerate the **Recovery Time Objectives** (**RTOs**) and **Recovery Processing Objectives** (**RPOs**), then local replication is cheaper.

In addition to these options, a bigger value for a **Processing Trigger Interval** can be used to control costs. Less frequent processing will result in slight delays in the arrival of data, but if the SLAs tolerate that delay, then it is absolutely justified.

## Handling latency trade-offs

Some events cannot tolerate the delay in traversing the entire multi-hop. In such cases, it is better to write to multiple sinks to facilitate the 'emergency' events to get across to reporting/notification apps. Each attempt to write can cause the output data to be recomputed, so it is important to cache the data before redirecting it to multiple sinks.

## Data reprocessing

There may be a need to reprocess data either because of a failed system coming back up after an outage or an unexpected deluge of data or a business logic change requiring a full re-computation of the already processed data. The following diagram lists the strategies for handling each scenario:

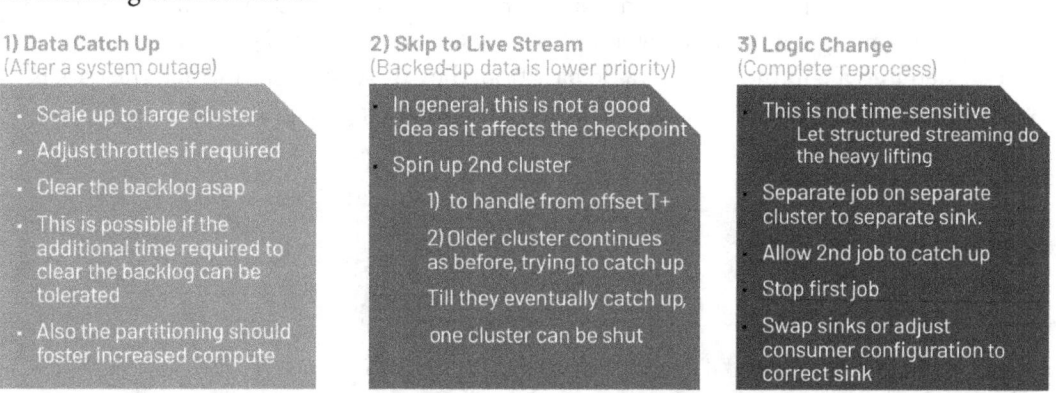

Figure 4.16 – Data reprocessing options

There are limitations on what changes are allowed to be made to a streaming query between restarts from the same checkpoint location (reference: *Recovering from failures with checkpointing*: `https://spark.apache.org/docs/latest/structured-streaming-programming-guide.html#recovering-from-failures-with-checkpointing`).

# Multi-tenancy

In a multi-tenant environment, the goal is to satisfy the client's need for data isolation. A typical strategy is to have a separate node per client or a separate job/cluster for each client. This, of course, leads to high costs and poor resource utilization, not to mention high maintenance across large numbers of jobs from both a code and DevOps perspective.

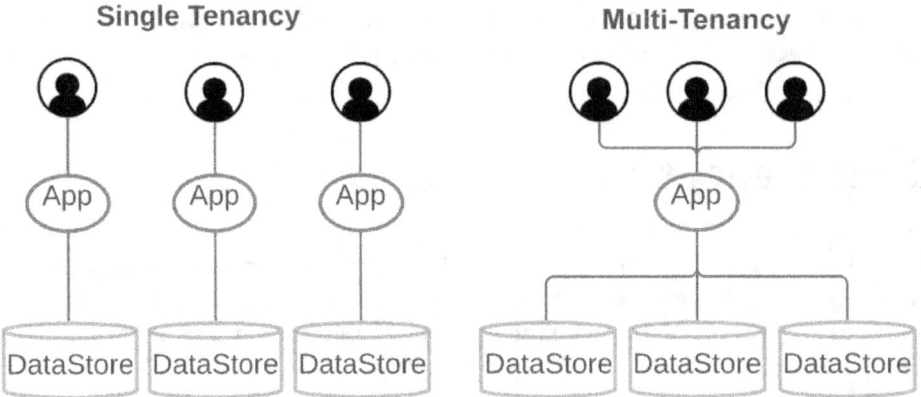

Figure 4.17 – Single tenancy versus multi-tenancy

These are some recommendations for designing a data platform, keeping the needs of multi-tenancy in mind:

- A common pattern for solving this is referred to as the **Mux-Demux pattern**, where all the incoming data initially leverages a single job/cluster for getting the data ingested into a common Delta table partitioned by the client from which the data is demuxed into separate groups, as shown in the following diagram:

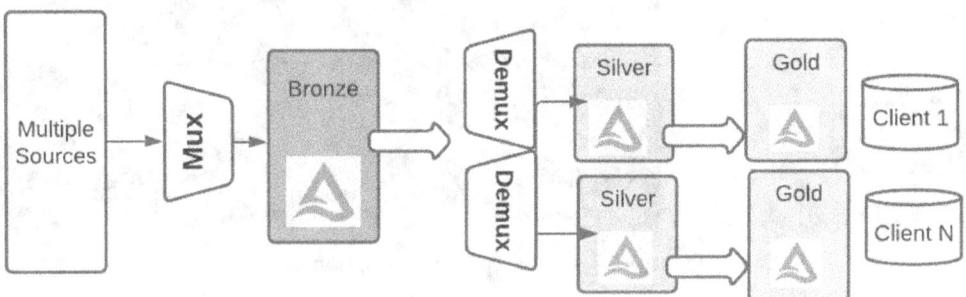

Figure 4.18 – The Mux-Demux pattern

- This will eventually cause an undue load on the driver and it might be necessary to assign multiple such clusters to balance the load.

- There can be separate datastores for each client in the serving layer where aggregated data is stored to feed the reporting dashboards.

- Separate views on the data can be created for individual clients who wish to see selected data in raw zones.

Multi-tenancy is an essential requirement of Software-as-a-Service (SaaS) offerings. Multiple clients will use the same software, so there are some efficiencies to be attained regarding reusing parts of the computational infrastructure as long as their data is well segregated.

# De-duplication

Given a tracking use case, the goal is to accurately report on the number of unique callers in a given reporting interval. The strategy is to use watermarking (to deal with permissible lateness of incoming data) and dropDuplicates:

```
uniqueVisitors =
    .withWatermark("event_time", "10 minutes")
    .dropDuplicates("event_time", "uid")
```

If not used properly, there are some negative consequences, so things to watch out for include the following:

- Incorrect use of watermarking.

- The size of the state store could eventually blow up, leading to **Out-Of-Memory (OOM)** errors.

- Edge error scenarios are not easily reproducible; it takes a while to reach the bad state, so stress testing helps to ferret it out faster.

- By default, Spark remembers all the windows forever and waits for all late events. For small data volumes, this is OK, but over time, resource utilization will spike.

These are some recommendations for handling data de-duplication properly:

- Include a watermarked timestamp column to prevent the state from growing unbounded.

- The number of partitions in the state store should never change for any stateful processing. So, repartitioning after a deduplication to rebalance the data is a good idea.

- The `foreachBatch` + `MERGE INTO` construct is slower but a preferable compromise to the deduplication results (where some late events were dropped).

Further reading on conditions for watermarking to clean aggregated state can be found here: `https://spark.apache.org/docs/latest/structured-streaming-programming-guide.html#conditions-for-watermarking-to-clean-aggregation-state`.

# Streaming best practices

Gartner has identified five levels of streaming capabilities to determine the maturity of an organization in its journey toward stream analytics as data is converted into information and optimized to extract every ounce of insight from it. It goes from presenting what happened, and why it happened, to what will happen. The five stages are as follows:

1. Ingesting the data

2. Orienting the individual lines of business to be data-aware

3. Using model capabilities to advise business support systems to help make the decisions through extended testing

4. Automation, culminating in a learning capability that can adapt to changes in data

5. Operating conditions

Here is a graphic representation of these five stages:

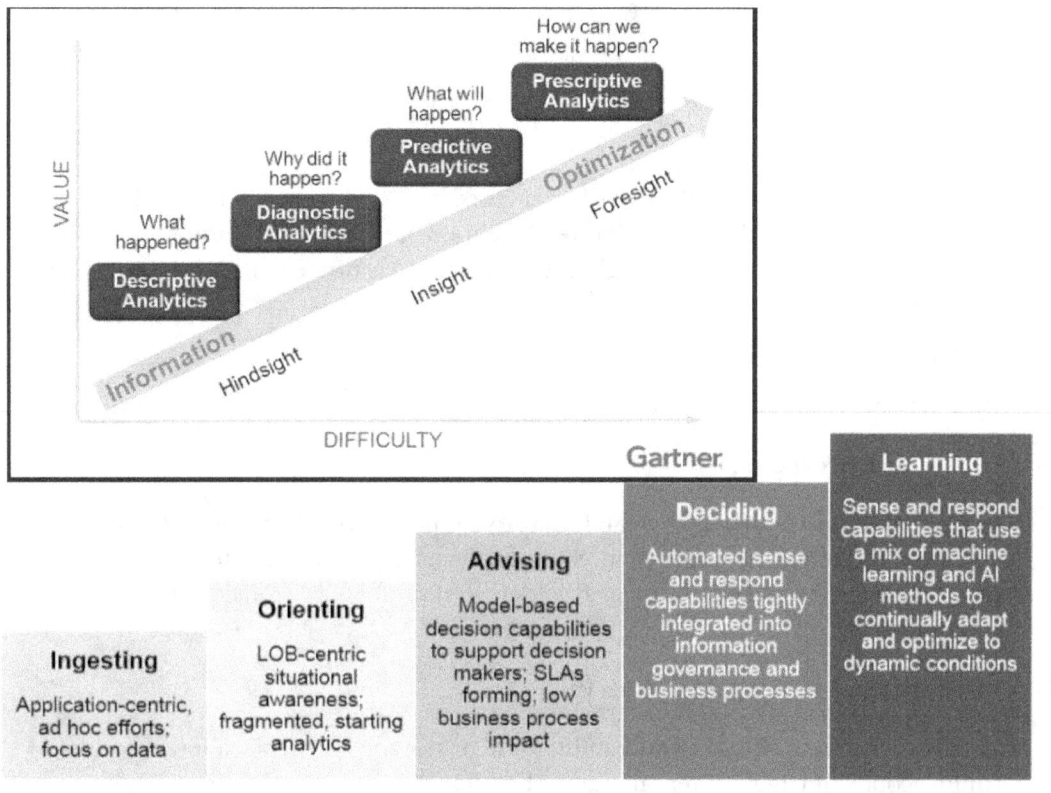

Figure 4.19 – Gartner's Streaming Analytics Maturity Model

**Reference**

https://blogs.gartner.com/nick-heudecker/five-
levels-of-streaming-analytics-maturity/

Here is a summary of some of the best practices for designing large-scale streaming pipelines:

- Understand data processing requirements and SLAs:

  - Ask the right questions regarding the speed of data arrival (**TPS: transactions per second**).

  - Understand end-to-end latency requirements. The knee-jerk reaction of business is to say ASAP, in which case ask how the data will be consumed and by whom. If the data is not consumed at a high frequency, there is no need to produce it with that frequency.

  - Get a ballpark cost estimate.

  - List existing technologies to understand integration points.

- Always use Checkpoint:

  - Later versions of Spark offer **AQE (Adaptive Query Execution)**, where runtime statistics influence the query plan. SQL shuffle partitions are created with every join/aggregate operation. If using an older version of Spark, consider tweaking `sql.shuffle.partitions` appropriately.

- Introduce a processing trigger interval:

  - To avoid frequent checking and an increase in storage API costs, a few seconds/milliseconds of trigger interval may be better.

- Optimized writes:

  - `Delta.autoOptimize.optimizeWrite = true` to reduce the number of files written.

  - To avoid re-computations when writing to multiple data sinks, we should cache the output DataFrame/dataset, write it to multiple locations, and then uncache it.

- Planning for projected capacity:

  - How many job definitions/runs/streams per cluster are anticipated?

  - Leverage the Mux-Demux architecture – where everything lands in bronze and is then broken down into separate clusters according to the tenant's needs.

In the next section, let's summarize what we have learned in this chapter.

## Summary

Information is dynamic and constantly evolving, which is why success in business is based on how we make use of this continuously changing data.

The modern data platform is built on business-centric value chains rather than IT-centric coding pipelines. The focus is to provide insights faster by turning event streams into analytics-ready data. Stream processing naturally fits with time series data and supports the detection of patterns over time. For some scenarios, streaming is a must, for example – sensor data, advertisement data, server security logs, and clickstream data. In some others, it is not, but every batch job can be regarded as a streaming job with a longer trigger interval and because the dial is configurable, it can be tweaked to make it more real time in line with business demands without having to rewrite the pipeline.

In this chapter, we focused on stream processing for ingesting, processing, and storing data. In the next chapter, we will look at how to take these various datasets and consolidate them in Delta Lake to break down the silos and democratize the data so that it is available readily to the parties that need it.

# 5
# Data Consolidation in Delta Lake

*"Faith makes you stable and steady. It brings out the totality in you.*

*Consolidation of your energy is faith. Dissemination of energy is doubt."*

*– Sri Sri Ravi Shankar, on Transcendental Meditation*

In the previous chapters, we discussed the quality of Delta and why it has become the first choice in big data processing. In this chapter, we will focus on how to consolidate disparate datasets into one or more data lakes backed by Delta so that you can build all kinds of use cases on a single source of truth without having to move data or stitch together multiple systems. We have already looked into the special features that Delta offers, including ACID transaction support, schema evolution, time travel, fine-grained data operations, and also big data design blueprints (such as the medallion architecture) in the context of data workflows. In this chapter, we will use those features to demonstrate the lakehouse architecture and the ease of creating and maintaining use cases on top of Delta Lake without compromising data quality and governance. Speed to insight is an important consideration for all businesses, and this new setup should help use cases to remain perpetually competitive and agile.

We will be covering the following topics in this chapter:

- Why consolidate disparate data types?
- Delta unifies all types of data
- Avoiding patches of data darkness
- Curating data in stages for analytics
- Ease of extending new use cases
- Data governance

# Technical requirements

To follow along this chapter, make sure you have the code and instructions as detailed in this GitHub location: `https://github.com/PacktPublishing/Simplifying-Data-Engineering-and-Analytics-with-Delta/tree/main/Chapter05`

Examples in this book cover some Databricks specific features to provide a complete view of capabilities. Newer features continue to be ported from Databricks to the Open Source Delta.

Let's get started!

# Why consolidate disparate data types?

Every finger on your hand serves a purpose, and although you could use them interchangeably to press a key or push the door, it cannot be denied that you need all of them when doing complex tasks such as typing, playing the piano, or grasping an object tightly. That is because the whole is usually greater than the sum of its parts. This is true in the data world as well. You may have specialized **Enterprise Resource Planning (ERP)** and **Customer Relationship Management (CRM)** operational data stores such as Salesforce, **Systems, Applications, and Products (SAP)**, Jira, or others. However, there is greater value in looking at a cross-section of your customers or accounts by bringing all the relevant and crucial pieces together to get the value without drowning in data. A lot of transactional systems may be structured (or at best semi-structured), but a lot of secondary data that is used to augment the core data is unstructured in nature, such as images of receipts, snippets of voice conversations in customer calls from call center logs, video snippets from surveillance cameras, and text from social media. This is where a data lake or a data mesh of mini data lakes divided by lines of business comes in handy.

There are connectors to specialized data stores such as Cassandra, Elastic, Redis, Redshift, SQL databases using JDBC, and different cloud storage systems. When so many sources of data are being pumped in, care needs to be taken to maintain data lineage so that repudiation of the source can be accounted for. Also, not all the data can be exposed to all the stakeholders. There have to be rules around who owns the data, who can view it, and who can manage it. Data along a pipeline is constantly being refined with transformations, normalizations, and joins with larger datasets, and it is important to expose these details to the consumer so they can decide at which point in the pipeline to consume the data. These details aid in data consolidation without compromising lineage and governance. This is where a good data lake comes to the rescue and provides for not only the ingesting piece but the transparency, lineage, and governance aspects.

The lifecycle in a data lake is different from the lifecycle of other operational stores in the sense that data assets typically live much longer and may be worked upon by different compute engines. We are talking about cloud ecosystems where storage is highly scalable, reliable, and reasonably affordable. Computing is the expensive part, and most modern platforms have a separation of compute and storage so that costs do not grow exponentially over time. Different industries have different requirements for how long data should be available for reporting purposes, and that is typically what drives the retention period after which data is moved to colder, cheaper storage or deleted altogether. The pipeline processes can vary in complexity, but follow the four phases of data retrieval, data processing, data analytics, and data storage, as shown in the following diagram:

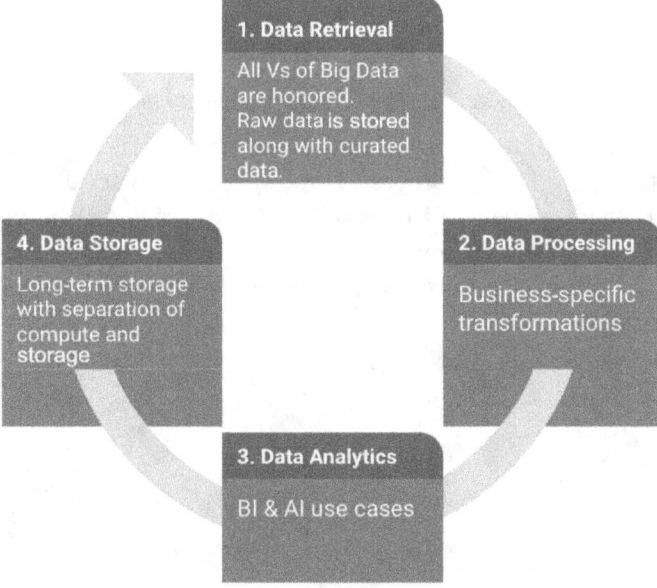

Figure 5.1 – Data lifecycle in Delta Lake

The diagram touches on the main dimensions and capabilities that a mature data lake offers as core capabilities. ETL deals with the data sources, and orchestration ensures that this process is done as part of an automated robust pipeline with dependencies, retries, and alerting capabilities. **Master Data Management (MDM)** manages schema, provides a central place for data discovery, and exposes data lineage so data personas can examine them to decide on the trustworthiness of the data sources they need for their use cases. Quality metrics should be continuously generated and reviewed for anomalies in data pattern trends so that they can be detected and corrected early on. Data access and security hooks allow privileged and role-based access to data, and governance allows admins to set the rules and monitor the audit logs. The following diagram shows the primary capabilities that Delta enables:

Figure 5.2 – Delta Lake capabilities

There are a lot of system integrator vendors active in this space offering drag and drop utilities to simplify the consolidation process. Some specialize in certain data sources or types of data, and others are generalists. Some examples are FiveTran, Informatica, StreamSets, Talend, and Syncsort. Some specialize in mainframe data, others specialize in streaming data, and yet others specialize in certain clouds or certain connectors.

# Delta unifies all types of data

In this section, we will give you some examples of how to ingest (read) different data types into a Spark DataFrame and save it all in the Delta format in the data lake using a common API, `df.write.format.("delta")` and this curated data is the single source of truth for all BI and AI use cases, as shown in the following diagram:

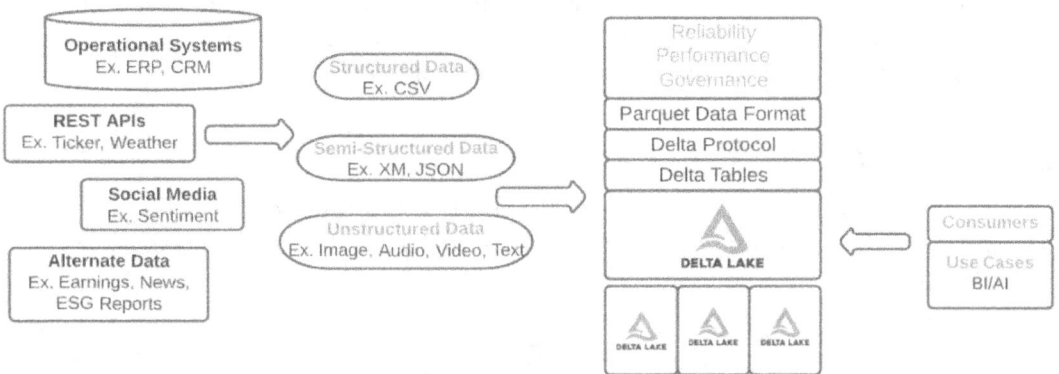

Figure 5.3 – Consolidating all your data in a central Delta Lake

The three main data types are **structured**, **semi-structured**, and **unstructured**, and Spark native APIs (along with vendor-aided connectors) can ingest data from a wide variety of data sources to create a curated view that all consumers can access, depending on their privilege levels. In some cases, different **Lines Of Business** (**LOBs**) may want to have their own mini data lake, and that is perfectly fine as it follows a hub-spoke model of a central data repository for common access and specialized ones with tighter guardrails for sensitive and domain-specific data resembling a data mesh. **Delta Sharing** is an open, secure protocol that can also be used to expose curated datasets to third parties. This will be addressed in *Chapter 11, Operationalizing Data & ML Pipelines*.

Let's consider a Customer360 use case in which a retailer needs to track customer engagement through various channels to provide the best product recommendations and coupon codes and prevent churn. Browsing history and prior sales transactions are examples of structured data. The location of the customer could lead to store suggestions. Weather, holidays, and vacation data are gleaned through REST APIs and are usually semi-structured in nature. Tweets from customers and reviews on Facebook with images and videos are examples of unstructured data. In the next section, we will look at each category in more detail and see how bringing it into a common Delta format can unify various cross-sections and hide the underlying complexity of the data producers.

## Structured data

Let's start with structured data, starting with the simplest **Comma-Separated Values (CSV)** format, with which you can either infer the schema or pass the schema explicitly. Please refer to the following code snippet:

```
#Either infer schema
df_csv = spark.read.format('csv')
        .options(header='true',inferSchema='true')
        .load('<path to csv data>')

# If the data types are not what you expect, you can use cast to convert,
# Ex. Convert colX to Integer data type, provided that is a permissible conversion
from pyspark.sql.types import IntegerType
df_csv = df_csv.withColumn("colX", df_csv["colX"].cast(IntegerType()))
```

When specifying the schema, there is always the possibility that the specified schema does not match some/all of the incoming data, in which case you need to specify how you would like unmatched data to be handled. The three options are as follows:

- PERMISSIVE (allow)
- DROPMALFORMED (drop bad records)
- FAILFAST (stop the pipeline)

The first is the default, and it allows you to examine the bad records. If you specify "_corrupt_record" to the schema, it will display a string with the original data. A mistake in the user-specified schema will cause any row with a mismatched data type value to be nullified.

```
# Explicitly specify the schema where say there are 3 incoming fields
from pyspark.sql.types import StructType, StructField, IntegerType, DateType

schema = StructType([ StructField("col_01", IntegerType()),
                      StructField("col_02", DateType()),
                      StructField("col_03", IntegerType()) ])

df_csv = spark.read.csv('<path to csv data>', header=True, schema=schema)
```

Avro and Parquet file formats are binary and hence are not human-readable. They are suitable for storing Structured data as they retain the structure/schema of the data and its data types in a more efficient manner.

## Semi-structured data

XML and JSON are examples of semi-structured data. Data and schema are self-describing via tags and can be nested. Attributes are not ordered (unlike in structured data) but they have start and end tags and can be parsed into rows. XML is the older of the two and is being replaced by JSON in several scenarios. There is a lot of semi-structured data because its extensible and free-style information exchange suits the internet. Data stores such as MongoDB and Couchbase and search engines such as Elastic favor JSON. Most REST interfaces are JSON based. Though it is human readable, it is sometimes not efficient because it is verbose and not splitable.

Use the spark-xml library to parse XML. You can validate individual rows against an XSD schema using rowValidationXSDPath. You can also use the from_xml method to parse XML in a string-valued column in an existing DataFrame. DataFrames and XML can be interconverted, and excludeAttribute can be used to control which attributes to ignore.

JSON data can also be in single or multiple lines. Multiline mode can be turned on as shown in the following example. from_json() and to_json() convert between JSON and strings into the struct/map type:

```
df_json = spark.read.json("<json data path>") \
               .options("multiline", "true")
```

Parsing is a necessary evil. Sometimes, data practitioners prefer to keep JSON strings as they are and parse them during querying.

## Unstructured data

The bulk of the data around us is unstructured. Image, audio, video, and text are examples of unstructured data, and each of them can be ingested and stored in Delta format for subsequent uniform query capabilities. Let's start with image data. On loading a DataFrame, there are additional metadata fields such as `origin`, `height`, `width`, `nchannels`, `mode`, and `data`.

```
df_img = spark.read.format("image") \
                .load("<path-to-image-data>")
```

There are some limitations of image data format, mostly relating to large sizes and limited decoding methods, which is why the `binaryFile` format is used more often.

```
df_bin = spark.read.format("binaryFile") \
        .option("mimeType", "image/*") \
        .load("<path-to-image-dir>")
```

For pure text, `spark.read.text("<path-to-image-dir>")` yields a data frame and `spark.read.textFile("<path-to-image-dir>")` yields a dataset.

What about video files? You can think of video as a series of image frames. Using either a video file reader such as FFmpeg, the video data can be read via a **User-Defined Function (UDF)** and exploded into frames, which can then be read using `binaryFile` on a distributed infrastructure using parallelism and persisted in Delta format. Drug manufacturers analyze video recordings of their clinical patients to determine drug efficacy. To be able to do this at scale and get the required data points for each patient without exposing **Protected Health Information (PHI)** is a fairly challenging task. It incorporates several disciplines, including object identification, pose elimination, and anonymization from every frame of the video.

Let's now move on to audio files. It may be necessary to subtitle a lecture or translate it. The `binaryFile` format can be used here. There are also several single node libraries, such as librosa, for audio processing that can be wrapped in a pandas UDF to process the

```
#Load wav data files using binaryFile format reader
df_au = spark.read.format("binaryFile") \
        .option("pathGlobFilter", "*.wav") \
        .option("recursiveFileLookup", "true") \
        .load("<data path to audio files>")
```

In the next section, we will scrutinize data quality and ensure that reliable data is available for the consumers no matter when they decide to tap into the data lake.

# Avoiding patches of data darkness

There are different lenses with which to measure data quality. In simple terms, you want clean, complete, accurate, consistent, timely, and unbiased data. You want your stakeholders to trust the data so they can build more sophisticated data products. Multiple personas using different views of the data should not get contradictory data points, and at no point should false facts be made visible because compliance and audit will uncover it sooner or later.

There are some common problems that every organization dealing with big data grapples with that lead to compromises in data quality, namely failed production jobs, lack of schema enforcement, lack of data consistency, lost data, and compliance requirements such as the GDPR. Let's examine these problems in the context of a simple airline use case of showing flight delays and see how Delta's features help address data quality.

## Addressing problems in flight status using Delta

This use case is to report on flight delays, and three sources of data are used, namely flight schedule data (which could be continuously streaming from Kafka or cloud storage), plane metadata (which is a slowly changing dimension and gets infrequent updates), and weather data (which is updated every 5 minutes and has ramifications on flight schedules).

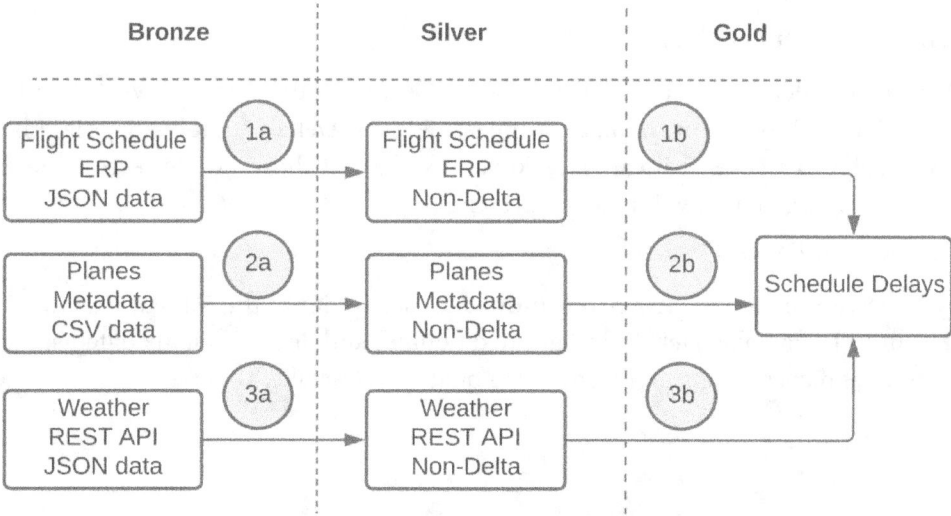

Figure 5.4 – Data flowing in an airline schedule reporting scenario

Let's look at each of the problems mentioned earlier:

- **Failed production job**

  Any of the bronze to silver or silver to gold pipelines can fail at any time on account of data issues, human error, infrastructure problems, and a host of other reasons. Let's say 3a fails. Then, the silver table for weather data is corrupt, which in turn corrupts the final gold table. Delta's ACID transaction support ensures no partial data is left in any target table and hence no damage is done, and the next run of the weather pipeline will re-establish business as usual.

- **Lack of schema enforcement**

  We redeployed an enhancement to the flights pipeline via a new job, and this accidentally introduced an incompatible schema change. Now, thousands of other schemas in the silver zone for flights are broken and the final gold table is unusable. Delta's schema enforcement and schema evolution help us control the metadata changes that are acceptable and safe.

- **Lack of consistency**

  We ran multiple jobs at the same time, of which some were `append-only` and other upstream jobs had `delete` with inconsistent data storage! Some pipelines had writes in progress, and the final gold table was left with partially completed data. Again, Delta's ACID transaction support, along with unified batch and streaming support, comes to the rescue. Reader/writer isolation properties ensure that partial data is never presented to end users.

- **Cannot recover lost data**

  A new job is deployed that accidentally specifies the mode to be "overwrite," and this deleted all the historical data for the last 5 years! Delta's audit history via time travel helps to recover from an occasional mistake and allows you to recreate data from the past given a version or a timestamp.

- **GDPR compliance**

  We receive a request to change one individual data point in the 5 TB of data in the entire Delta Lake. Delta's fine-grained updates and deletes help fix data using updates and corrections to improve data quality in the lake.

Each of the previous scenarios could have easily turned into a catastrophe, but with Delta, we can address them quite simply. Delta can be both a sink and a source for streams where data arrives continuously, in frequent batches (such as hourly) or in infrequent batches where it may change every couple of hours or days or more. In the next section, we will look at how domain knowledge can be used while building pipelines to put additional constraints and expectations around the data.

# Augmenting domain knowledge constraints to quality

Dirty data is frustrating for both internal stakeholders such as data scientists and BI analysts and is also damaging when used by external stakeholders because wrong predictions could lead to business and reputation losses. Sometimes a lot of domain knowledge is required to do justice to the data adjustments where generic cleansing rules may do more damage than good. The Harvard Business Review has estimated data cleansing alone costs a whopping $3 trillion (reference: `https://hbr.org/2016/09/bad-data-costs-the-u-s-3-trillion-per-year`). Delta provides the pieces you need to build your own quality pipeline because each business has unique needs, and having the quality pipeline close to the data teams allows quick remediation and tweaks to heal data pipelines more quickly.

Some common data scrutiny includes checks for missing or incomplete data, differing formats or data types (otherwise known as schema mismatch), and user errors when writing data producers and transformers. These bad records may be on account of files that are unreadable, or maybe the files themselves were processable but the resulting data did not meet our expectations. Spark offers three parse modes, namely PERMISSIVE (which allows the bad record but creates a default column called _corrupt_record where the offending data is placed for future inspection; DROPMALFORMED (which just ignores all the bad records), and FAILFAST where the entire pipeline is halted the minute the first offending data element is encountered. The following snippet demonstrates how to use it.

```
data = """{"a": 1, "b":2, "c":3}|{"a": 1, "b":2, "c":3}|{"a": 1, "b, "c":10}""".split('|')

corruptDF = (spark.read.option("mode", "DROPMALFORMED").json(sc.parallelize(data)))
corruptDF = (spark.read.option("mode", "PERMISSIVE")
                         .option("Corrupt", "_corrupt_record").json(sc.parallelize(data)))

corruptDF = (spark.read.option("badRecordsPath",  myBadRecords).json(sc.parallelize(data)))
```

Another common scenario is the need to drop duplicate records. One user performs a write operation on Delta table A. At the same time, another user performs an append operation on Delta table A. In a subsequent silver table, this can be detected and eliminated as shown here:

```
duplicateDedupedDF = duplicateDF.dropDuplicates(["id",
"favorite_color"])
```

Yet another scenario is a need to impute missing data. Strategies for imputing missing values include the following:

- **Dropping these records**: This results in data loss and works only when you do not need to use the information for downstream workloads:

  ```
  corruptDroppedDF = corruptDF.dropna("any")
  ```

- **Adding a placeholder**: This allows you to see missing data later on without violating a schema.

- **Basic imputing**: This allows you to have a good guess at what the data could have been, often by using the mean of the data that is present:

  ```
  corruptDroppedDF = corruptDF.na.fill({"temperature": 68,
  "wind": 6})
  ```

- **Advanced imputing**: You can use advanced strategies such as clustering machine learning algorithms for imputation.

Adding constraints at the **Data Definition Language** (**DDL**) layer is the best way to handle quality at the onset without waiting for others to layer in quality checks. Some of the fields can be generated from existing fields as well. The following example shows the creation of a partitioned Delta table:

```
CREATE TABLE default.people10m (
  id INT,
  gender STRING,
  birthDate TIMESTAMP,
  dateOfBirth DATE GENERATED ALWAYS AS (CAST(birthDate AS DATE)),
  salary INT
)
USING DELTA
PARTITIONED BY (dateOfBirth);

ALTER TABLE default.people10m
ADD CONSTRAINT dateWithinRange CHECK (birthDate > '1900-01-01')
```

Some tools/libraries, such as Great Expectations AWS Deequ, that provide simple APIs and are built for big data can be used alongside Delta to define data constraints and expectations and capture data metrics as a parallel pipeline that is being monitored for outliers. Yet another approach is to embed quality right into the pipeline as done by Delta Live Tables, currently a proprietary offering from Databricks. In the next section, we will see why continuous monitoring is needed to ensure that quality standards remain high.

## Continuous quality monitoring

Data quality is an ongoing battle. Building your pipeline and demonstrating the end-to-end flow does not mean everything is golden. There will be times when your dashboards may look sparse, and your reports will be off because data did not come through for that day and the charts went off the trendline, not because the data is all bad but because there was a little patch of darkness and your pipeline is falling behind, or there is no data refresh, or perhaps there is a bug somewhere. These blips cause a lot of commotion and can be expensive to deal with. Many times, an engineer has taken the fall for these issues, but the term *good pipeline but bad data* was coined to deal with data issues. Constant monitoring of relevant metrics is the only way to detect anomalies well in advance before they turn into real issues.

This is not to be confused with *dark data*, which Gartner defines as data collected by a business to adhere to compliance reasons but is not leveraged as part of the analytics and BI of the organization. It is estimated that 90% of a firm's data could be dark data, which is why there are sometimes blind spots in an organization as there is potential for insights to be harnessed. Dark data assessment is a scan of a company's unstructured data to report on the potential value of that data.

# Curating data in stages for analytics

The raw data has to be wrangled and transformed to be consumable and ready for analytics. Each data persona may look at different data aspects or features, and there is no reason for all of them to run repeatable cleansing functions because if they did, they could all have multiple copies of data and unnecessary processing cycles, which is both time-consuming and expensive. This is where a good data catalog and design blueprints help to maintain discipline, offer data discovery opportunities for reusable components, and prevent redundant work. We have already looked at the medallion architecture, and the bronze, silver, and gold zones are where data is forged and made usable.

## RDD, DataFrames, and datasets

This is a good time to refresh concepts around RDD, DataFrames, and datasets. **RDD** stands for **Resilient Distributed Data** and is the original low-level construct of Spark. RDDs have to be optimized at each stage and cannot infer schema. DataFrames are a higher-level construct and deal with rows and columns of data and address RDD problems; however, they do not offer compile-time type checking. Datasets are strongly typed objects with schema mapping and are available only for JVM languages, that is, Java and Scala.

## Spark transformations and actions

Spark RDD operations are of two types: transformations and actions. Transformations are lazy, meaning that they are not applied until an action API is encountered. The immutable RDD lineage is captured in a **Directed Acyclic Graph** (**DAG**) and so there is no danger of losing anything because the steps to recreate are known. Transformations can be narrow or wide. In narrow transformations, all the elements of a partition remain as such, whereas in a wide transformation, they get shuffled. Examples of narrow transformations are `map()` and `filter()`, whereas examples of wide transformations are `groupByKey()` and `reduceByKey()`.

# Spark APIs and UDFs

Spark offers a lot of APIs in multiple languages that are tested for performance and scale and should be the first choice to explore when looking for data transformation operations. Only if one does not exist should you create your own **User-Defined Function** (**UDF**). In the industry, there are several similar-sounding terms with subtle differences:

- **User-Defined Function** (**UDF**): Works on a row and produces a row, also known as one:one mapping

- **User-Defined Aggregate Functions** (**UDAF**): Works on several rows and produces a row, also known as many:one mapping (examples are sum and count)

- **User-Defined Transformation Functions** (**UDTF**): Works on multiple rows to produce multiple rows, also known as many:many mapping (an example is explode)

Built-in APIs are the most efficient, so use them whenever possible because we would miss some Spark optimizations if we were to code some of them ourselves. Also, edge cases, null value handling, and a lot of checking have to be carefully handled. However, there are cases where the business logic requires custom code and we have no choice but to deal with it explicitly. Python UDFs have an overhead with serialization and deserialization that JVM-based UDFs are immune from. So, care should be taken to use pandas UDFs when working with Python.

UDFs allow multiple column inputs, and complex outputs can be designated with the use of a defined schema encapsulated in a StructType() object. The following is an example of how to define a pandas UDF by adding the annotation @pandas_udf to a function:

```
from pyspark.sql.functions import pandas_udf, PandasUDFType

@pandas_udf('double', PandasUDFType.SCALAR)
def pandas_plus_one(v):
    return v + 1
```

The following is an example of how to call the UDF by name and pass in the required columns:

```python
from pyspark.sql.functions import col, rand

df = spark.range(0, 10 * 1000 * 1000)
df.withColumn('id_transformed', pandas_plus_one("id")).show()
```

When writing a regular Python UDF (which is not recommended), you need to register it:

```
<name of UDF> = udf(<name of python function>, <optional return type>)
```

UDFs can be called from SQL code as well, but the way to register them is a little different:

```
spark.udf.register(<name of UDF>, <name of python function>, <optional return type>)
```

In this section, we went over the various data transformations to refine raw data to make it ready for consumption. In the next section, we will see how Delta helps to future-proof our tech investment by making it easy to onboard new use cases and providing a nimble and agile path to production.

# Ease of extending to existing and new use cases

Business demands are constantly changing and evolving, and a well-designed data lake provides scalability not just for the storage and compute needs of existing use cases but also in supporting new processing modes to meet emerging unseen scenarios. In other words, it is agnostic of the specifics of the industry vertical or use case. Exploratory data analysis, machine learning, and business analytics can all work from the same view of the data. New tools and frameworks can be layered in, and as long as the data does not have to move around, it provides a solid base for data personas to focus on their specific use case goals instead of having to solve common infrastructure-level issues over and over again, increasing their productivity and the agility of use cases getting to production. Delta is an open format and, once curated, it is available to a host of other tools using connectors. The following section talks about some popular connectors.

# Delta Lake connectors

Although developed by Databricks, the Delta Lake project was contributed to by the Linux Foundation in 2019, making it an independent open source project and not controlled by any single company. The `delta-io/connectors` repo is a central place where you can access a lot of Delta connectors to read and write to data in the Delta Lake using a host of other tools in the ecosystem. Some of the main ones are listed here:

- A manifest file containing the list of data files to read for querying a Delta table is a common approach for querying data in a Delta lake. Data warehouses such as Redshift and Snowflake and query engines such as Presto and Athena use this approach.

- BI tools such as Power BI have native support for Delta, and `sql-delta-import` is a utility for importing data from a JDBC source into a Delta Lake table.

- The Hive connector uses an uber JAR to read data in Delta lake and `kafka-delta-ingest` is used for streaming data from Kafka.

- The DBT plugin supports drag and drop functionality and connects to Spark clusters hosted on EMR, Databricks, or Docker (using ODBC, HTTP, Thrift, and Spark session).

- The Delta Rust API provides low-level access to Delta tables and is intended to be used with data processing frameworks such as `datafusion`, `ballista`, and `rust-dataframe`, and it can also provide native bindings in Python, Ruby, and Golang.

- Delta Standalone Reader is a JVM library for reading Delta Lake tables. Unlike `https://github.com/delta-io/delta`, this project doesn't use Spark to read tables and it has only a few transitive dependencies. It can be used by any application that cannot use a Spark cluster.

In the next section, we will look at how different industry verticals have nuanced variations of their respective Delta Lakes.

# Specialized Delta Lakes by industry

Delta helps improve *data fidelity* by storing the original raw data alongside the processed data for data personas to compare, reprocess, and re-evaluate some of the business processing and fix it instead of overwriting data and erasing all means of recovering from an inadvertent error. We saw in *Chapter 2, Data Modeling and ETL,* that data modeling exercises are needed even for big data. There are benefits of both *schema on read* and *schema on write*, depending on your position in the pipeline, and instead of having to choose one over the other, Delta provides a schema evolution mechanism to effectively adapt to changes in data patterns without compromising the quality of the data or being completely surprised by the amount of change happening without your consent. Building permission-based and role-based access control on top of datasets puts guardrails around data access and use. This is how Delta offers better *data flexibility and manageability* options. Time travel capabilities capture operations by time and version alongside the user details, providing full *traceability and auditability* of the transformations performed on the data. It offers multilingual APIs to cater to the skillset of a wider audience that may need to work on a SQL engine one day and move to a machine learning engine the next or streaming the next. Different compute engines can work on the data because not only are the compute and storage separate, but the storage is in *open format*. Let's look at some specialized data lakes by industry and the kinds of datasets they consolidate.

## Healthcare and life sciences Delta Lake

Clinical data lakes provide a wealth of information on a wide variety of use cases, from recommending your fitness plan to exploring drug efficacy for clinical trials to genome sequencing and much more. Various risk models can be developed that munge patient-generated data with their master health record, compare it to the history of similar patients to identify risk factors, and address issues before they become fatal. Other than patient health and treatment, the secondary use cases that a clinical data lake can support include predicting healthcare costs and evidence-based care.

Healthcare generates a lot of varied data from the electronic health records of patients collected through various channels, such as pharmacy records, electronic patient health records, wearables, claims, trials, medical imaging, and registries. In fact, it is estimated that a single patient produces roughly 80 MB of medical data annually. So, scale is the first challenge. Delta, backed by Spark's distributed architecture, is a perfect fit for the growing data in this sector. A lot of this data is unstructured, so it needs to be on a modern data lake capable of handling imaging data, doctors' transcribed notes, ultrasound recordings, and so on.

A lot of time is spent gathering and gluing all the disparate sources linking to a patient, and this is where Delta Lake, with its unified batch and streaming and support for all data types, can come in handy. Wearable devices such as insulin regulators and heart monitors send data quickly, and the underlying data platform needs real-time capabilities for both ingesting and analyzing the data quickly. Delta's support for z-order allows data scientists to query data across multiple dimensions. Data generated from expensive medical devices taking scans and blood reports can now be analyzed by multiple algorithms and linked with other data points of a patient to get a better understanding of their health condition and suggest suitable treatment. Delta's ACID semantics and stream processing makes it possible to analyze critical patient data in-stream. All of this sensitive data has to follow **Health Insurance Portability and Accountability Act (HIPAA)** `https://www.cdc.gov/mmwr/preview/mmwrhtml/m2e411a1.htm)` regulation guidelines in the United States, and **Electronic Healthcare Records (EHR)** systems have to comply with interoperability guidelines and require the consent of the patient. The ability to support fine-grained updates allows PHI data masking.

The ability to organize all health-related data at scale, providing real-time insights while honoring data quality and compliance using the lakehouse architecture on a low-cost architecture is what is needed in the face of rising healthcare costs.

## Industry 4.0 manufacturing Delta Lake

Manufacturing covers a lot of sub-domains, including supply chain, predictive maintenance, warehouse robots, production processes, fulfillment logistics, and cost optimizations, all of which benefit from analytics and AI. The IoT era popularized smart factories and digital twins, and this has resulted in the rapid generation of data that needs to be analyzed quickly.

According to the world economic forum, the oil and gas industry could increase its production severalfold by investing in drilling locations predicted by machine learning algorithms using data collected from sensors, thereby helping the environment. This is an example of the data in the data lake being as valuable as the oil itself: `http://reports.weforum.org/digital-transformation/oil-and-gas-on-the-cusp-of-a-digitally-fuelled-new-era/`.

## Financial services Delta Lake

Financial services cover use cases across the banking, insurance, and capital markets domains. The main focus is to minimize and mitigate risk, provide audit and compliance reporting, personalize services for better customer experience, and provide real-time insights to foster quicker data-driven decision-making. Transaction analysis for fraud and risk management use cases including **Know Your Customer** (KYC) and **Anti Money Laundering** (AML) relies on the data to be as complete and accurate as possible.

Delta's streaming capabilities help provide the foundation for scalable time-series processing, catering to a wide variety of analytics use cases. Joining transactional data with a wide range of unstructured alternate data is possible because of Delta's support for unstructured and semi-structured data. Delta sharing is especially handy for compliance and regulatory reporting requirements.

## Retail Delta Lake

Businesses need to provide sufficient differentiators to improve brand loyalty and prevent customer churn. The focus is to get products closer to customers with personalized recommendations and rewards to keep them engaged and loyal. Businesses need a 360-degree view of their customers and their own business to do so. There is a need to develop operational efficiencies and stronger collaboration with manufacturers, distributors, and partners in the value chain.

Delta's support for real-time ingestion and analytics helps provide better **Point Of Sale** (POS) and customer behavior segmentation analysis. Timely data results in better demand forecasting, which leads to accurate safety stock optimizations and on-shelf availability of popular products.

In this section, we examined different lakes that industry verticals adapt for their specific use cases by creating reusable components that increase the speed to production of new use cases. In the next section, we will look at how the inherent capabilities of Delta help with governance requirements.

# Data governance

Data democratization and self-service capabilities are some of the advantages of data lakes. A data governance layer is imperative to put the right guardrails in place while allowing stakeholders to get the most business value from the generated and curated data and insights. A good data catalog is essential for producing actionable insights in any data-driven organization. Cloud vendors have their own offerings, such as AWS Glue, Azure Purview, and Azure Data Catalog. Apache Atlas is probably the most popular open source offering, and there are vendors who specialize in this area such as Alation and Collibra.

The three primary goals of governance are the following:

- Keeping data secure and only the right privileges and roles dictate access to data
- Ensuring the quality of the stored data is high so that it is meaningful to its consumers, who then develop trust in their data and hence the insights generated on top of the data
- Discovering data so that there is healthy reuse of the data in the way it was intended

Data comes in at a very fast pace and should be sorted right from the get-go to avoid breaches and exposures. So, integrating with the development and operational parts of the business is important. A framework with clear guidelines on data modeling practices and pipeline development for new data should guide data personas. Centralizing the metadata helps its administration. Ownership is granted to the dataset authors and curators. Lineage details add to the credibility of the dataset. Authentication and authorization rules define who has access to data and what actions they can perform. Finally, data retention closes the loop on the lifecycle management of the data asset. Let's look at a few scenarios.

# GDPR and CCPA compliance

GDPR and CCPA are compliance requirements that most enterprises have to adhere to. Any individual can request updates to their data records, and the company is obliged to not only honor the request but also provide proof of executing the desired operation. The first part of compliance is to understand the data subject rights such as what personal information is collected; the ability to erase, update, or export personal information; and the ability to respond to such requests within the statutory timelines. A lack of structure makes it challenging to locate individual records. Most data lakes do not have full support for fine-grained updates and deletes. Petabyte-scale data is a hindrance to finding the records of a few individuals in a bunch of tables in a timely manner. This applies to all historical datasets as well. Ad hoc queries for a full table scan to find individual records are expensive and time-consuming, and the impact of this could result in significant penalties and loss of credibility. Let's see how Delta can address each of these tasks:

- Search for the personal information in a performant and fast manner:

  - Delta performance is aided by features such as auto-indexing, caching, and compaction.

- Reliably change, erase, or export the identified personal information:

  - Fine-grained updates and deletes along with ACID transaction guarantees ensure the requests are committed to storage.

  - Delta Sharing allows secure sharing of data.

- Store information in a non-identifiable form:

  - Pseudo anonymization is the process of using lookups for the reversible tokenization of personal identifiers. This is where Delta's ability to support fine-grained updates comes in handy.

- Report and audit the requested changes:

  - Transaction logs enable reporting on changes to data for the audit and security review process.

  - Delta Time Travel provides an easy way to view these data modifications.

The history of a Delta transaction log, that is, the various versions of data over time, can be viewed using the `DESCRIBE HISTORY <delta path or table>` command, and the output would look something like this:

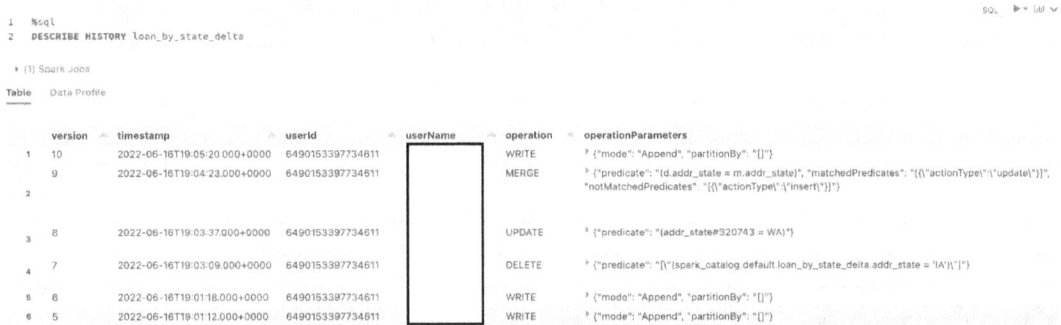

In the next section, we will see how to set privileges for data access and manipulation based on the user's role.

## Role-based data access

There will always be multiple data personas and automated non-user accounts, or service principals, working on a data product. Managing access, roles, and responsibilities along with usage is a necessary job function of an admin. You can enable table access control to grant, deny, and revolve access to data in databases, tables, views, and functions per user or group. Either the whole data artifact or a subset of it can be guarded. Let's consider an example where an admin owns some shared and sensitive tables in the Payroll database. They then provide another user the privileges to just read the shared but not the sensitive data:

```
GRANT SELECT on TABLE Payroll.Shared to `user@company.org`;
DENY ALL on TABLE Payroll.Sensitive to `user@company.org`;
```

Some datasets are more sensitive than others, depending on the use case. In any case, avoid putting personal data in the open. Raw data can be pseudo-anonymized using a lookup table.

Configuring access control to data files and database artifacts is the first part. The second is to monitor via audit logs to ensure that those rules are honored and any violation or tampering is recorded and reported.

All this results in efficient operations, better security and compliance, and reduced expenditure.

# Summary

It is interesting to note how the term *data lake* came about. It is not called a pond as a pond is perceived to be small. It is not called a sea or ocean because the saltwater makes it look murky and the waves are rough and uncontrolled. It is not called a stream as "streaming" is already heavily used in the context of real-time processing. It is not a river because water drains off, whereas the vision of a data lake is that of a pristine reservoir of water that provides food and shelter to a lot of flora and fauna and could turn into a swamp if you're not careful with governance and management. In this chapter, we went over the need for data consolidation and how Delta helps with data reliability, quality, and governance, giving us curated analytics-ready data and preventing silos and swamps. Data, once curated, remains in an open format and is used in multiple use cases by different data personas, enabling them to be more agile in on-boarding new use cases and maintaining existing ones.

It is fair to say that a data lake provides storage and compute for processing and storing big data and its associated metadata at scale. Data remains in an open format, and other tools of the ecosystem continue to tap into the single source of truth. Unlike data warehouses, data lakes store raw data alongside refined curated data. So, use cases with different SLA needs can tap into different views of the data as well as track the lineage of transformations. Data lifecycle management, where older data is archived or dropped completely, is at the discretion of the compliance rules in the specific industry vertical. Diverse AI and BI use cases can all be built on top of a data lake. It can optionally push curated data to other serving layers or share it with approved third-party vendors and partners. Data catalogs maintain schema and business definitions along with access and ownership rules. Delta makes it easy to achieve all these capabilities by providing the foundations of a protocol and data format that brings the capabilities of a **Massively Parallel Processing** (**MPP**) data warehouse right into a data lake. In the next chapter, we will explore various data operations that will be performed on data stored in a data lake.

# 6

# Solving Common Data Pattern Scenarios with Delta

*"Without changing our pattern of thought, we will not be able to solve the problems we created with our current pattern of thoughts"*

*– Albert Einstein*

In the previous chapters, we established the foundation of Delta and how it helps to consolidate disparate datasets, and how it offers a wide array of tools to slice and dice data using unified processing and storage APIs. We examined basic **Create, Retrieve, Update, Delete** (CRUD) operations using Delta and time travel capabilities to rewind to a different view of data at a previous point in time for rollback capabilities. We used Delta to showcase functionality around fine-grained updates and deletes to data and the handling of late-arriving data. It may arise on account of a technical glitch upstream or a human error. We demonstrated the ability to adapt to changing schema. This happens when the underlying application is upgraded and new fields are introduced, as a few of the old ones may have changed, or maybe one or two of them have been dropped because they are no longer considered useful. Delta simplifies all these operations and now we will leverage them to build bigger reusable patterns.

There are common data operations that we have been addressing in different ways for a long time, but with every tech advancement, we adopt a different approach to solving the same problems to squeeze in an extra ounce of performance and efficiency. In this chapter, we will create reusable code building blocks to tackle everyday data processing needs so that these modular patterns can be applied in a wide variety of use cases across different industry verticals over and over again. In a small application, or while doing a **Proof of Concept** (**POC**), it is, of course, very easy to rip and replace. But for a large-scale enterprise data pipeline, it has to be done very carefully as the ramifications are widespread. The analogy is that of maneuvering a speed boat versus a large ship. Moving data is always error-prone, so once it's curated, we will look at strategies to minimize data movement. There are some cases where it is necessary to create a backup or a clone of the curated data in a different region/location and it is important to understand the distinction between these scenarios. Adding data to the lake is always simpler than managing changes to data at scale. In this chapter, we will look into solving some of these problems with Delta, namely the following:

- Understanding use case requirements
- Minimizing data movement with Delta time travel
- Data cloning
- Handling **Change Data Capture** (**CDC**)
- Handling **Slowly Changing Dimensions** (**SCD**)

# Technical requirements

To follow the instructions of this chapter, make sure you have the code and instructions as detailed in this GitHub location:

```
https://github.com/PacktPublishing/Simplifying-Data-
Engineering-and-Analytics-with-Delta/tree/main/Chapter06
```

Examples in this book cover some Databricks-specific features to provide a complete view of capabilities. New features continue to be ported from Databricks to the open source Delta.

Let's get started!

# Understanding use case requirements

Each problem that a client brings up will always have some similarities to a problem you may have seen before and yet have some nuances to it that make it a little different. So, before rushing to reuse a solution, you need to understand the requirements and the priorities so that they can be handled in the order of importance that the client values them. A good way to look at requirements is by demarcating the functional ones from the non-functional ones. Functional requirements specify what the system should do, whereas non-functional requirements describe how the system will perform. For example, we may be able to perform fine-grained deletes from the enterprise data lake for a GDPR compliance requirement, but it takes two days and two engineers to do so at the end of each month, so it will not meet the requirements of a 12-hour SLA. The technical capabilities exist, but the solution is still not usable. The following diagram helps you classify the requirements into these two groups as they will influence your choice of trade-offs as you build the solution.

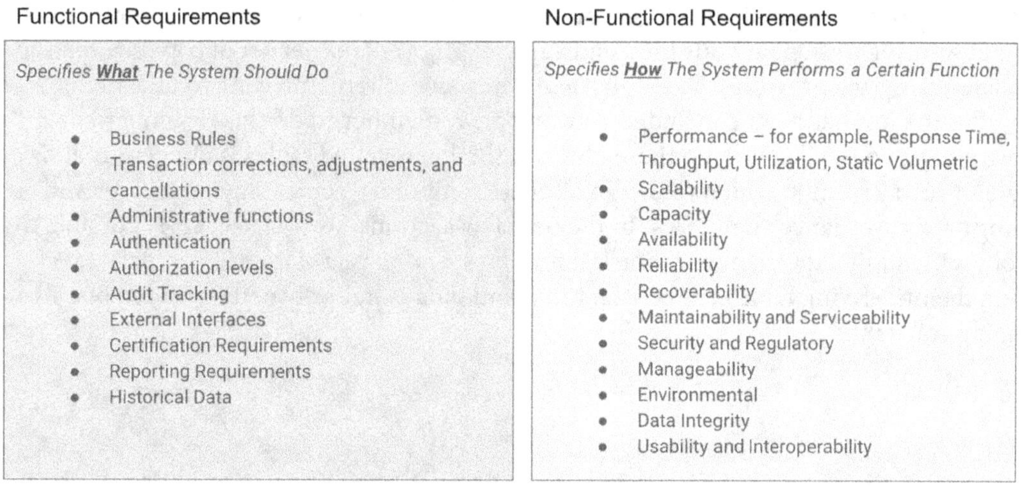

**Functional Requirements**

*Specifies **What** The System Should Do*

- Business Rules
- Transaction corrections, adjustments, and cancellations
- Administrative functions
- Authentication
- Authorization levels
- Audit Tracking
- External Interfaces
- Certification Requirements
- Reporting Requirements
- Historical Data

**Non-Functional Requirements**

*Specifies **How** The System Performs a Certain Function*

- Performance – for example, Response Time, Throughput, Utilization, Static Volumetric
- Scalability
- Capacity
- Availability
- Reliability
- Recoverability
- Maintainability and Serviceability
- Security and Regulatory
- Manageability
- Environmental
- Data Integrity
- Usability and Interoperability

Figure 6.1 – Functional and non-functional requirements

In *Chapter 5, Data Consolidation in Delta Lake*, we emphasized the need to keep data as a **single source of truth** (**SSOT**) and build multiple disparate use cases on a single view of the data; this is because every time you move data, there are plenty of opportunities to make errors and compromise the quality of data, in addition to the additional cost and operational burden. In the next section, we will first look at a few techniques to minimize data movement and also explore scenarios where it is absolutely necessary.

# Minimizing data movement with Delta time travel

Apart from ensuring data quality, the other advantage of minimizing data movement is that it reduces the costs associated with data. To prevent fragile disparate systems from being stitched together, the first core requirement is to keep data in an open format for multiple tools of the ecosystem to handle, which is what Delta architectures promote.

There are some scenarios where a data professional needs to make copies of an underlying dataset. For example, to make a series of A/B tests in the context of debugging and integration testing, a data engineer needs a point-in-time reference to a data snapshot to compare for debugging and integration testing purposes. A BI analyst may need to run different reports off the same data to run some audit checks. Similarly, an ML practitioner may need a consistent dataset because experiments have to be compared across different ML model architectures or against different hyperparameter combinations to make a model selection. Before anyone realizes, too many copies have been made and it is difficult to keep track of the original version. Another person decides to collaborate in your efforts and, not wanting to step on your toes, ends up creating yet another set of copies. Creating new datasets means providing access to these datasets and remembering to take them away after the test has been concluded. Few people remember to do that, leading to data exposure situations. Cleaning up these files and their associated tables is dangerous and people shy away from it, which results in all these temporary copies staying forever and causing data governance challenges. In the next few sections, we will see if it is possible to do rollbacks, audit data changes in the last few days, and have a static view of data, even though the underlying table is in constant flux and data is not only getting added but also changing constantly.

Every data operation on a Delta file or table format automatically gets a timestamp and version number. Users can then use `'timestampAsOf'` or `'versionAsOf'` in Scala, Python, or SQL to look at a snapshot of the data:

- Python syntax:

```python
# using timestampAsOf
df = spark.read.format('delta').option('timestampAsOf', '2022-01-01').load('<path to delta table>')

# using 'versionAsOf'
df = spark.read.format('delta').option('versionAsOf', '54').load('<path to delta table>')
# Alternate way of specifying
df = spark.read.format('delta').load('<path to delta table@v54>')=


# These can also be exposed as utility functions
def loadAsOfData(dataPath) :
  return spark.read.format('delta').load(dataPath)

# and called
df = loadAsOfData('<path to delta table@v54>')
```

- SQL syntax:

```sql
%sql
-- using timestampAsOf
SELECT * from <delta_table> TIMESTAMP AS OF '2022-01-01'

-- previous day
SELECT * from <delta_table> TIMESTAMP AS OF date_sub(currentt_date(), 1)

-- from a more granular time
SELECT * from <delta_table> TIMESTAMP AS OF '2022-01-01 05:05:00.555'

-- Three ways of using versionAsOf
SELECT * from <delta_table> VERSION AS OF 54
SELECT * from <delta_table>@v54
SELECT * from delta.'<path to delta data>@v54'
```

Changes can be audited by looking at the history of the Delta table, as demonstrated here:

```
DESCRIBE HISTORY <delta table>
Or
DESCRIBE HISTORY delta.'<path to delta data>'
```

Apart from other metadata details, such as the user, the number of files affected, and the operation performed (such as an insert/update/delete/merge), the syntax highlights the version and timestamp values that we were using in our previous commands. When you select * from a Delta table, it is retrieving data from the latest version. To find the latest version number or timestamp, you can use the following:

```
sql_s = "SELECT max(version) FROM (DESCRIBE HISTORY <delta
table>)"
version = spark.sql(sql_s).collect()
```

Delta time travel can help fix an inadvertent operational error where some data accidentally got deleted or updated, as shown in the following examples:

```
%sql
INSERT INTO <delta table>
SELECT * FROM <delta table> TIMESTAMP AS OF date_sub(current_date(), 1)
WHERE <condition>
```

The following is an example of merging changes when a condition is matched:

```
MERGE INTO <delta table> target
USING <delta table> TIMESTAMP AS OF date_sub(current_date(), 1) source
ON source.userId = target.userId
WHEN MATCHED THEN UPDATE SET *
```

An ML practitioner may collaborate with others to run a series of experiments for two weeks on a given dataset, and they all need to know the version of data to use for their base training, test, or validation datasets. This will help them compare results fairly and justify their choice of the winning model.

A BI analyst is always looking for patterns and trends in data. Temporal data analysis to see how many new sales were made in the last day or week, for instance, can be easily done using the same time travel capabilities of data without requiring the exact day/time values, which makes it possible to automate such workloads.

```
%sql
SELECT
count(distinct transactions) -
(SELECT count(distinct transactions) FROM <delta table>
        TIMESTAMP AS OF date_sub(current_date(), 1))
FROM <delta table>
```

In the next section, we will look at the cloning functionality of Delta, which has various applications for data engineers and data scientists and analysts alike by giving them the ability to back up and refresh data across environments or reproduce an experiment with fidelity on a copy of the data in a different region. It is important to distinguish between time travel and cloning capabilities. In time travel, you are rewinding to a different version or timestamp of the dataset, whereas in cloning, you are making data available in a different setup or environment.

# Delta cloning

**Cloning** is the process of making a copy. In the previous section, we started out by saying that we should try to minimize data movement and data copies whenever possible because there will always be a lot of effort required to keep things in sync and reconcile data. However, there are some cases where it is inevitable for business requirements. For example, there may be a scenario for data archiving, trying to reproduce an ML flow experiment in a different environment, short-term experimental runs on production data, the need to share data with a different LOB, or maybe the need to tweak a few table properties without affecting original source especially if there are consumers leveraging it with some assumptions.

**Shallow cloning** refers to copying metadata and **deep cloning** refers to copying both metadata and data. If shallow cloning suffices, it should be preferred as it is light and inexpensive, whereas deep cloning is a more involved process.

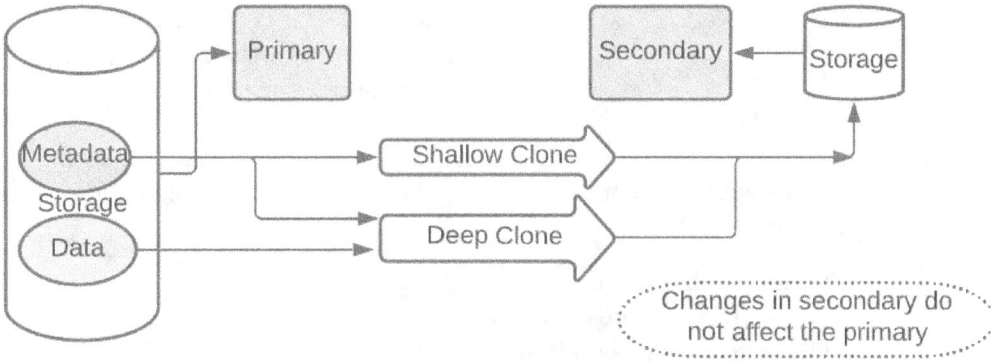

Figure 6.2 – Shallow versus deep cloning in Delta

The previous figure shows the two paths for copying data and metadata for the two scenarios of shallow and deep cloning between a primary and a secondary datastore. It should be noted that after the clone operation, changes to the secondary do not affect the primary. The following table highlights the differences even further with guidance on when to use each.

| | Delta Shallow Clone | Delta Deep Clone |
|---|---|---|
| When to use | You want to try out a numberof changes and not muddy the Delta history. | You want complete isolation, for instance, in a disaster recovery scenario. More expensive to create. |
| How to create | CREATE Table <delta shallow clone> SHALLOW CLONE <orig delta table> LOCATION 'new data dir for shallow clone' | CREATE Table <delta deep clone> DEEP CLONE <orig delta table> LOCATION 'new data dir for deep clone' |
| Original table | Has many history transaction log files. | Has a lot of history transaction log files. |
| After clone | A new Dir appears but no data is copied. All metadata around schema, partition is copied but not table properties. | A new directory appears and the latest version of data is copied across. All metadata around schema, partition is copied but not table properties. |
| Run vacuum on original | FileNotFoundException will be thrown. Run the CREATE and REPLACE commands again to recreate the shallow clone. | No change to the deep clone table. |

| After changes to new table | Original table data files remain as before. New data files appear in the shallow clone folder.<br><br>The cloned table has an independent history from its source table. | Original table data files remain as before. New data files appear in a shallow clone folder.<br><br>The cloned table has an independent history from its source table. |
|---|---|---|
| After changes to original table | Original table data files change.<br><br>No change reflected in new<br><br>input_filename() refers to the files from the time the clone was created. | Original table data files change.<br><br>No change reflected in new<br><br>On running the deep clone again, the incremental changes get across to new location. |

Figure 6.3 – Shallow versus deep cloning

You can combine cloning with time travel capabilities to copy a certain snapshot of data. In the next section, we will look at two closely related operations – **Change Data Capture (CDC)** and **Slowly Changing Dimensions (SCD)** – one addresses the flow of data from the perspective of the source, the other from the perspective of the target table.

# Handling CDC

CDC is a process that identifies the classification of incoming records in real-time to determine which ones are brand new, which ones are modifications of existing data, and which ones are requests for deletes. Operational data stores are capturing transactions in OLTP systems continuously and streaming them across to OLAP systems. These two data systems need to be kept in sync to reconcile data and keep data fidelity. It is like a replay of the operations but on a different system.

# CDC

This is the flow of data from the OLTP system into an OLAP system, typically the first landing zone, which is referred to as the bronze layer in the medallion architecture. Several tools, such as GoldenGate from Oracle or PowerGate from Informatica, support the generation of change sets, or they could be generated by other relational stores that capture this information on a data modification trigger. Moreover, this could be an omni-channel scenario where the same type of data is coming from different data sources and needs to be merged into a common target table. These could be coming in a continuous stream or ad hoc occasional updates. Usually, the amount of incoming data is small when compared to the target tables they are intended for. Also, there may be numerous change datasets. The following diagram provides a reference architecture of change datasets landing at a defined refresh interval, getting ingested into the staging table, and finally being merged into the target table. Delta's ACID properties are critical to CDC/CDF use cases as it puts locks on the table and eliminates the need to worry about partial reads/writes.

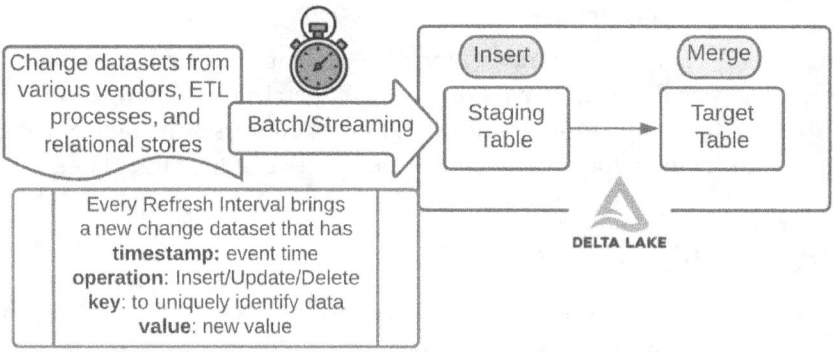

Figure 6.4 – CDC flow for change datasets

Among other things, the change dataset holds at least four pieces of critical information, including the timestamp of the event, the type of operation such as insert/update/delete, the unique key that can be used to join with a target table to determine whether it is an existing record or not, and the new value. There are two variations to the architecture:

- The first approach involves a temporary staging table where the change dataset lands as an insert followed by an insert overwrite type operation into the target table. For example, a continuously streaming setup, insert operation from a cloud storage location, or Kafka topic. The inner condition ensures we pick the latest record if multiples come. If it is a delete, then it is suppressed altogether so the staging table closely matches an SCD Type 2 table, and the final table resembles an SCD Type 1 table. We will cover SCD in the *Slowly Changing Dimensions (SCD)* section of this chapter.

- The second approach involves an INSERT or INSERT OVERWRITE into the staging table with a partition on the date/time field, followed by a MERGE INTO for the target table.

INSERT OVERWRITE is used to replace contents either for a particular partition or the entire table. MERGE is a more selective operation where new rows are added, and existing rows are updated or deleted based on a key match. The former is used when the bulk of the data in the target table has changed whereas the latter is used when there are fewer changes and only fewer rows are affected

Additional partition columns on date/time fields help ensure that operations are efficient and work off only the relevant datasets instead of doing full table scans, which, for a large table, is a very expensive operation:

```sql
%sql
-- First insert into a staging table
INSERT INTO staging_delta PARTITION(partition1)
SELECT * FROM incremental_view;

-- Subsequent insert into a target table
INSERT OVERWRITE target_delta PARTITION(partition1)
    SELECT *   FROM (
      SELECT A.*,
        RANK() OVER (PARTITION BY key ORDER BY event_timesamp DESC) AS RANK
        FROM staging_delta A.* WHERE <conditions are met>
    ) B
WHERE B.RANK = 1 AND B.FLAG < > 'Delete';

-- Before Consuming from the final target table, run optimize
OPTIMIZE T_FINAL ZORDER BY (col1, col2)
where date_col > current_timestamp() - INTERVAL 1 day;

SELECT * FROM T_FINAL WHERE COL1 = val and COL2 = val WHERE partition1='some value';
```

Whenever possible, it is a good idea to use partition columns for partition pruning and Z-order on columns used in the 'where' clause of consumption queries for better performance by the query engine from the target tables. The 'VACUUM' operation on the target table at periodic intervals helps to purge the older versions (as in copies of the data). On the staging table, a simple overwrite from the target table can help control additional data bloat so that all records where the rank is greater than one are removed.

If CDC is attempted on a large number of tables, it is recommended to use a generic pipeline with a configuration-driven approach to scale out to the different datasets, which may vary by parameters such as event name, timestamp, matching key field, partition column, and so on. A simple schema of table name, partition column, rank criteria, and strategy (insert versus update) can help to abstract out the code and configuration aspects and go a long way with the debugging and maintenance of the numerous pipelines. Any improvement to the base code of the pipeline will benefit all the pipelines.

## Change Data Feed (CDF)

This is the continuation of the flow of CDC data from the landing zone Delta tables downstream to curated silver and aggregated gold zones. It is worth pointing out that time travel and data versioning are done at the file level and not at the individual row level. CDF addresses issues around quality control and inefficiencies in a big data ecosystem. Every new ingestion round brings in new data and handling these row-level changes across dataset versions is hard to maintain. Efficiency is paramount as we are talking about the context of real-time systems, and it is hard to justify the time taken to scan whole datasets to understand which rows need to be merged.

CDF can be enabled at a cluster or table level. Additionally, it can be enabled on a brand new table (during the CREATE operation or on an existing table as a subsequent UPDATE). Once enabled, this feature is forward-looking, meaning previous ingestion setups are treated as is. The following example shows its use during a new table creation:

```sql
%sql
-- create bronze, silver and gold tables with deltaChangeDataFeed enabled
CREATE TABLE <delta_table>
( fields ….)
USING DELTA
TBLPROPERTIES(deltaChangeDataFeed = true);
```

The changes can be queried using the `table_changes` function either by stating a starting version or a starting and ending version as follows:

```sql
%sql
-- Initial load into bronze
INSERT INTO bronze_delta TABLE bronze_dataset;

-- Initial load into silver
INSERT INTO silver_delta
SELECT * FROM table_changes('bronze_delta', 1);

-- Initial load into gold
INSERT INTO gold_delta
SELECT * FROM silver_delta
INNER JOIN (SELECT DISTINCT … FROM table_changes('silver_delta', 1)) AS silver_cdf
 ON silver_delta.date = silver_cdf.date  AND silver_delta.some_key = silver_cdf.same_key
GROUP BY silver_delta.date, silver_delta.some_key;
```

An inner join of the silver zone table with the `table_changes` data helps identify the changed rows and aggregation on those rows for gold zone tables can now be done selectively to improve the efficiency of operations:

```sql
%sql
-- Subsequent load into bronze
INSERT INTO bronze_delta TABLE bronze_new_dataset;

-- Subsequent load into silver
MERGE INTO silver_delta
USING
 (SELECT * FROM table_changes('bronze_delta', 2)) AS bronze_cdf
 ON bronze_cdf.date = silver_delta.date AND
    bronze_cdf.some_key = silver_delta.same_key
WHEN MATCHED THEN
 UPDATE SET silver_delta.some_value = bronze_cdf.same_value
WHEN NOT MATCHED
 THEN INSERT (...) VALUES (...)

-- Subsequent load into gold
MERGE INTO gold_delta
USING
 (SELECT *  FROM silver_delta
 INNER JOIN (SELECT DISTINCT … FROM table_changes('silver_delta', 2)) AS silver_cdf
   ON silver_delta.date = silver_cdf.date AND silver_delta.some_key = silver_cdf.same_key
GROUP BY silver_delta.date, silver_delta.some_key) as silver_cdf_agg
ON silver_cdf_agg.date = gold_delta.date AND silver_cdf_agg.some_key = gold_delta.same_key
WHEN MATCHED THEN
 UPDATE SET gold_delta.some_value = silver_cdf_agg.same_value
WHEN NOT MATCHED
 THEN INSERT (...) VALUES (...)
```

It is worth noting that CDF operations, though highly optimized, are nevertheless an additional step and there are scenarios where they are a better fit than others. For example, if data coming in is always additive, resulting in an append-only mode, then additional checks of CDF are not really necessary. If most of the data is updated on every operation, then a direct `MERGE` operation suffices. If the source system has curated data in CDC format, then it is a perfect fit for the CDF data pattern, especially when these changes need to be transferred to downstream tables. The kill-and-fill scenario where entire data dumps are replaced is obviously not a fit.

To summarize, CDF makes handling data changes in Delta tables easy and efficient. The audit trail that it provides by itself is a huge benefit and it can be saved to cloud storage for future analysis. The ultimate use case is materialized views that provide real-time updates to BI dashboards by efficiently processing only the changed rows, instead of reprocessing entire tables every time a change comes through. In addition to Delta tables, streaming services such as Kafka and other data stores can also be a target for these change sets. In the next section, we will discuss a related concept – **Slowly Changing Dimensions (SCD)** – which is viewed from the target table's perspective.

# Handling Slowly Changing Dimensions (SCD)

Operational data makes its way into OLAP systems that comprise fact and dimension tables. The facts change frequently and are usually additive in nature. The dimensions do not change as often but they do experience some change, hence the name "slowly changing dimensions."

Business rules dictate how this change is to be handled and the various types of SCD operations reflect this. The following table lists them.

| SCD Type | What is it? | Notes |
|----------|-------------|-------|
| Type 0 | Fixed dimension - no change allowed | Too rigid |
| Type 1 | Only the latest value is retained, meaning the old value is overwritten | **Pro**: No additional memory/storage is required<br>**Con**: We cannot trace the history of modifications |
| Type 2 | Additional records are created for each change | **Pro**: You can trace back to the history<br>**Con**: The memory/storage is getting consumed due to keeping old records |
| Type 3 | Change handled by adding a new column | **Pro**: better solution to SCD 1<br>**Con**: Not sustainable over time and some systems may have a max limit on the number of columns |
| Type 4 | Uses historical tables | Similar to 3 in the sense that it not sustainable |
| Type 6 | Combines the approaches of 1, 2, and 3 | Overly complex |

Figure 6.5 – SCD types

Of all these alternatives, types 1 and 2 are the most popular in the industry. In the next section, we will explore them in more detail.

# SCD Type 1

This is fairly straightforward as there is no need to store the historical data; the newer data just overwrites the older data. Delta's MERGE constructs come in handy. There is an initial full load of the data. New data is inserted, existing data is updated, and deletes remove the data altogether.

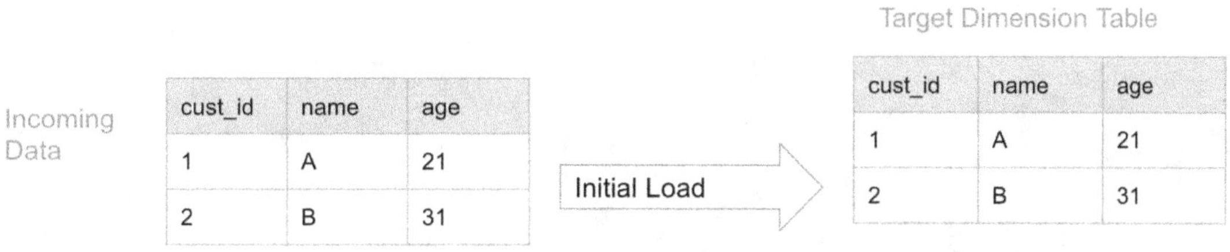

Figure 6.6 – SCD 1 – initial load

This is a visualization of the state of the target dimension table on the initial load of data the very first time we perform an ingestion operation. The subsequent operations are more interesting because decisions have to be made regarding which rows will be additions and which rows will require modification.

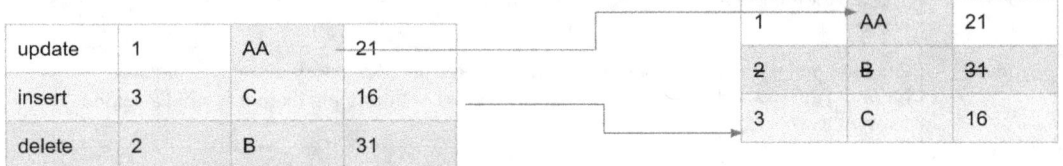

Figure 6.7 – SCD 1 – handling new data and changes to old data

In the following example, we are picking up new data from the bronze table (`src`) and merging into the target table (`tgt`) by joining user identifier, only if there has been a change in the `city` field. If a matching record in the target table is found, an update operation is done; otherwise, it is regarded as a new record that gets inserted. Joining on `user_id` is a must, whereas adding the city check is an optimization. If every day you were getting the same data dump, then the target table would be left untouched. A count of the number of rows updated, inserted, or deleted will make this point clearer.

```sql
%sql
MERGE INTO <target_delta_table_scd_1> tgt
USING
(SELECT * FROM <user_bronze_tbl> WHERE update_date = current_date()) as src
ON tgt.user_id = src.user_id
WHEN MATCHED AND tgt.city <> src.city
   THEN UPDATE SET *
WHEN NOT MATCHED
   THEN INSERT *
```

In the next section, we will discuss SCD Type 2, which is a very popular pattern as it retains older history.

# SCD Type 2

Much like in SCD Type 1, there is an initial load and any new data that comes later is added as new records. However, unlike SCD Type 1, any change to existing data is manifested as a new row. New columns around effective dates and recency are added to manage the change. Start and end dates to indicate the validity period of data help when going back in time. A current Boolean field helps to indicate whether that data is the most recent or not. All of this should be done as a single operation in order not to introduce race conditions, and this is where Delta's ACID properties come in handy.

The ability to do "as of" analytics is powerful. It allows you to rewind in time to recreate the view of data and understand exactly what the state of the world was then. Let's look at an initial load and subsequent ingestion to a dimension table.

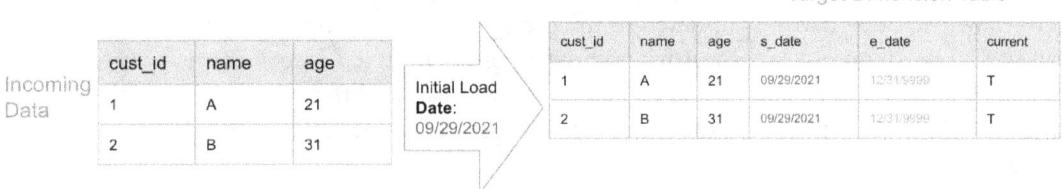

Figure 6.8 – SCD 2 – Initial load

As in SCD Type1, the very first ingestion is straightforward, all the rows are inserted as is, but do take note of the additional date fields and a Boolean field to indicate whether it is currently active or not. As expected, everything is set to `True`. On subsequent ingestions, the start and end dates get adjusted, and the older values are set to `False` to indicate that they are no longer current.

| Update | 1 | AA | 21 | Next Load Date: 10/2/2021 | 1 | A | 21 | 09/29/2021 | 10/1/2021 | F |
| Insert | 3 | C | 16 | | 2 | B | 31 | 09/29/2021 | 10/1/2021 | F |
| Delete | 2 | B | 31 | | 1 | AA | 21 | 10/2/2021 | 12/31/9999 | T |
| | | | | | 3 | C | 16 | 10/2/2021 | 12/31/9999 | T |

Figure 6.9 – SCD 2 - Subsequent load

We looked at star schema in the previous chapters when visualizing a fact table with several dimension tables around it. SCD Type 2 is a classic data warehouse problem that can be solved elegantly using Delta. The relationships between the fact and dimension tables remain unchanged; what changes, though, is the validity period of data in the dimension table as it relates to the "as of" state of the world and is controlled by the date/time ranges. At any time, there is only a single active or current record of the dimensional data to avoid confusion.

The same example that we used for SCD Type 1 is used again to demonstrate SCD Type 2. There are several ways to do this, and the one demonstrated uses a temporary src_changes, which in turn is a union of two other temporary tables, src_bronze_table and inserts_for_matched_changes. Now, user_id is used to create a new field called merge_key. Because the changed records are duplicated, a second set with a null merge_key is also added. The subsequent merge operation treats the match records as the ones whose end_date is updated to update_date, and current_indicator is set to false or an old record. A brand new record is inserted on a 'when not matched' whose current_indicator is set to true and the end date is left as null. Some variations of SCD Type 2 may use a very large time value for this end date. Both approaches are fine:

```sql
%sql
CREATE OR REPLACE TEMPORARY VIEW src_changes
AS
WITH
  src_bronze_table as
      (SELECT *  FROM <user_bronze_tbl> WHERE update_date = current_date()),
  inserts_for_matched_changes as
    (SELECT src.* FROM src_bronze_table src
    JOIN <target_delta_table_scd_2> tgt
      -- only update if city changes
      ON tgt.current_indicator = true and src.user_id = tgt.user_id and tgt.city <> src.city )

SELECT *, user_id as merge_key  FROM src_bronze_table
UNION ALL
SELECT *, null as merge_key FROM inserts_for_matched_changes;

MERGE INTO <target_delta_table_scd_2> tgt
USING src_changes as src
ON tgt.user_id = src.merge_key AND tgt.current_indicator = true
-- only city changes
  WHEN MATCHED AND tgt.city <> src.city
     THEN UPDATE SET record_end_dt = src.update_date, current_indicator = fals
  WHEN NOT MATCHED
     THEN INSERT (user_id, <rest of fields>, record_start_dt, record_end_dt, current_indicator)
           VALUES (user_id, <test of values>, update_date, null, true)
```

SCD Type 2 may seem long-winded, but once adopted and implemented, it is boilerplate code that can be reused for multiple tables by building a simple configuration-driven framework to cater to additional tables.

## Summary

Delta Lake with ACID transactions makes it much easier to reliably perform UPDATE and DELETE operations. Delta introduces the MERGE  INTO operator to perform Upsert/ Merge actions as atomic operations along with time travel features to provide rewind capabilities on Delta Lake tables. Cloning, CDC, and SCD are patterns found in several use cases that build upon these base operations. In this chapter, we have looked at these common data patterns and shown how Delta continues to provide efficient, robust, and elegant solutions to simplify the everyday work scenarios of a data persona, allowing them to focus on the use case at hand.

In the next chapter, we will look at data warehouse use cases and see if all of them can be accommodated in the context of a data lake. We will reflect on whether there is a better architecture strategy to consider instead of just shunting between warehouses and lakes.

# 7

# Delta for Data Warehouse Use Cases

*"It is not the strongest of the species that survives, nor the most intelligent. It is the one that is the most adaptable to change."*

*– Charles Darwin, On the Origin of Species. "Descent with modification"*

In the previous chapters, we went over the main capabilities of Delta and its edge over other data formats and protocols. Delta has its origins in data lakes, and we examined how Delta addresses the common challenges of traditional data lakes. In fact, it is the evolutionary next stage of lakes and fits into a new category known as the **data lakehouse**. So, it is a no-brainer that it is the preferred choice for any new data lake initiative, but what about data warehouse use cases? Is that a separate category of data scenarios that data lakes do not tackle effectively?

The data warehouse is a concept that was born in the 1980s! It was popularized by relational database platforms. As is true of every form of tech evolution, we will look at the evolution path for these platforms. The names for most other tech innovations of that era have been replaced with newer terminology, but we do still hear the term **data warehouse**. In this chapter, we will look at what a data warehouse really comprises, what specific use cases it solves, and which data personas it appeals to. We will also look at the other side of the coin and see which use cases it does not cater to, which personas cannot use it effectively, and the tech that it lacks. We will then look at Delta capabilities to see how it addresses each of these to become the de facto standard for all data use cases. Do modern data architectures have a place in a data warehouse box? What is the next step in the evolution of data warehouses? We will re-examine the lakehouse category in the context of the evolutionary trajectory of data lakes and data warehouses.

The following topics are covered in this chapter:

- Choosing the right architecture
- In which scenarios is a data warehouse useful
- Discovering when a data lake does not suffice
- Addressing concurrency and latency requirements with Delta
- Visualizing data using BI reporting
- Analyzing tradeoffs in a push versus pull data flow
- Considerations around data governance
- The rise of the lakehouse category

Remember: in the end, the only thing that truly matters in any tech is the business value that it brings to its users. Can access to the right data be democratized to all the data personas so they can do something useful with it to add value in a timely and cost-effective manner and give their business a competitive edge? Is the lakehouse category the holy grail? Is it easier for a data lake or a data warehouse to evolve into a lakehouse? Let's find out the winner of this tug of war.

# Technical requirements

To follow along this chapter, make sure you have the code and instructions as detailed in this GitHub location:

```
https://github.com/PacktPublishing/Simplifying-Data-
Engineering-and-Analytics-with-Delta/tree/main/Chapter07
```

Examples in this book cover some Databricks-specific features to provide a complete view of capabilities. Newer features continue to be ported from Databricks to the open source Delta.

Let's get started!

# Choosing the right architecture

If you were evaluating technologies to choose an architecture framework for your data use cases, what guiding criteria would you use to make the decision? The two popular ones out there are data warehouses and data lakes. The three main dimensions to consider are storage, compute, and governance.

The questions you should ask are: what are your use cases? If they are only BI, do you need to future-proof your investment? What kinds of data are you going to ingest? If it is mostly structured data, then do you want to consider the cost of migrating architectures if your stakeholders suddenly want to understand free text or image data? What about the existing skill set of the data personas who are going to work on this data platform? If you want to retain the best talent, you'll need to give them opportunities to learn and grow, so you'll need to spend some time understanding the capabilities offered by warehouses and lakes and not make a short-term decision. Once you decide on an architecture, you'll need to find out all the offerings in that space and compare them so that you can make a good business investment that will stand the test of time. The following diagram shows the basic building blocks in the context of a warehouse and a lake.

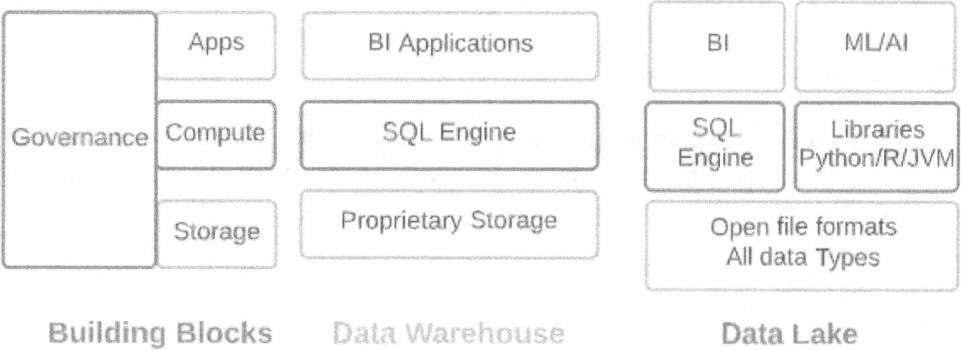

Figure 7.1 – Deciding on the right architecture – data warehouse, data lake, or both?

As you can see, data warehouses aim to solve BI use cases, and data lakes add support for ML along with unstructured data. But what about a data governance block? In the next few sections, we will examine the areas that each cater to in more detail.

There is a popular quote by John Owen, a theologian: "*Data is what you need to do analytics. Information is what you need to do business.*" This reminds us that data accumulation is a means to an end; converting data to information is what adds value.

# Understanding what a data warehouse really solves

At its core, a data warehouse is a data repository of disparate data sources, mostly *structured* to help answer BI analysis questions using *reports and dashboards*. It is typically used by heads of departments and businesses to get a bird's-eye view of how the organization is doing holistically using SQL interfaces. It analyzes operational data to predict growth and identify bottlenecks and other business stragglers to help the business evaluate its performance using KPI metrics in the face of its competitors and plan more strategically. The older on-premises offerings are moving into the cloud to take advantage of the elasticity of cloud computing.

This can be simplified into two main parts:

- Base underlying storage:

  - Cloud storage is increasingly popular on account of it being affordable, scalable, and reliable.

- The analytic layer built on top of storage:

  - The analytic layer houses several pieces beyond just data, such as the metadata, schema, data model, taxonomies, lineage, access privileges, KPIs, metrics, and others. The terms "BI" and "warehouse" are almost synonymous.

Enterprises invest in their BI strategies to drive sales, improve revenue, understand trends and bottlenecks, and speed up the decision-making process to maintain their competitive edge, as shown in the following diagram:

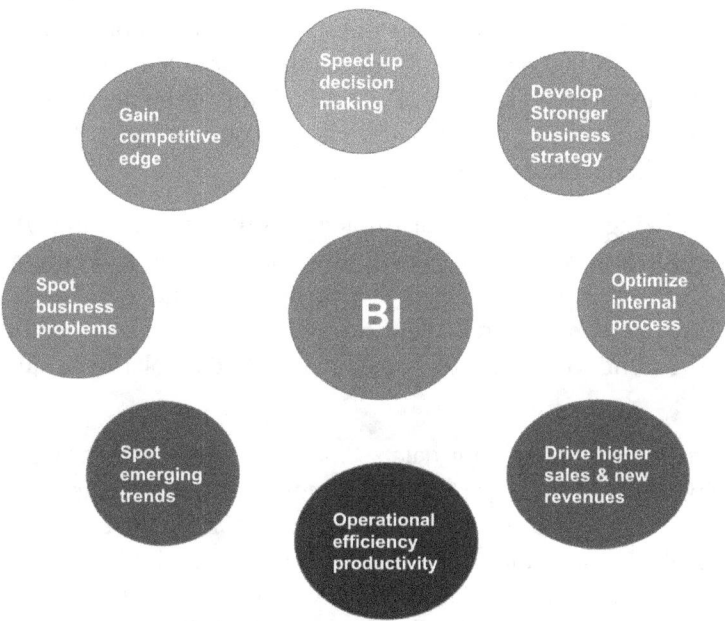

Figure 7.2 – Benefits of a BI system

There is a subtle difference between **Business Intelligence** (**BI**) and **Business Analytics** (**BA**) systems. The former tries to address the "*What and how,*" while the latter addresses the "*Why and what next.*" All of this sounds great, but now let's examine some aspects that are not so palatable.

## Lacunas of data warehouses

Irrespective of where it is hosted, at its heart, a data warehouse's data is *proprietary*, which gives it an inherent, unfair advantage over open source because the warehouse can change the format to its advantage. In other words, multiple tools of the ecosystem cannot use the data as it is, nor can you just decide to shut down your existing warehouse in favor of a newer, shinier one on the market. Yes, that's right, you are trapped in a vendor lock-in zone. Even though you know there is a better thing out there, you have to protect your earlier investment because migrating from one warehouse to another is expensive, time-consuming, and fraught with all kinds of risks, as is expected when multiple vendors have their own implementation nuances for the same data. You wonder why? After all, it exposes data using just SQL, which brings us to the next problem.

**Database Administrators** (**DBAs**) are a key persona and are regarded as the keepers of a warehouse. One thing to watch out for is the separation of compute from storage. Traditionally, they have been tightly coupled, which means that as the volume of data increases, you need beefier server nodes, and you are paying for it to be on all the time irrespective of whether you actually use some of the older data or not. This is why DBAs are careful to limit the historical retention duration; that is, older data would be dropped, impacting your ability to do trend analysis effectively. Some of the newer warehouses, such as Snowflake, recognized this problem and have decoupled compute from storage, but by and large, the other vendors have not.

The main persona working in warehouse technology should be skilled in SQL. It also requires a lot of administration, and DBAs are indispensable, revered figures that keep the lights running. These DBAs can't just be handpicked from the market, they need to be trained on the specific warehouse and its nuances. In other words, some of those skills are non-transferable. The trend toward cloud-hosted managed data platforms takes some of this pressure off.

Your only tool to wrestle with all of your data scenarios and use cases is SQL! All the software engineering skills and power of high-level modern languages is thrown out the window. SQL is great, but what about using the right tool for the job at hand? These are no options on the table. If you're lucky, the SQL you are expected to use will be ANSI-compliant so you don't have to relearn proprietary variants, but more often there will be variations to get used to. For example, Teradata uses **Basic Teradata Query** (**BTEQ**). IBM Netezza SQL is its own variant of SQL that runs on the Netezza data warehouse appliance. Greenplum Database PL/pgSQL is based on Postgres PL/pgSQL, which is a subset of Oracle PL/SQL. There are lots of built-in functions that you can leverage, but what happens when you want to implement your own logic? Some of them offer procedures and user-defined functions as alternate hooks, but what happens when you want to create or reuse more complex algorithms? Another downside of a warehouse is the inability to import libraries easily. This brings us to the next problem: being left out of ML opportunities.

Today's data platforms need to offer the ability to do both AI and BI use cases. Data Warehouses have served the need for BI use cases but fall short for AI ones. Today, every business is leveraging AI to get ahead of its competitors, imagine being stuck with just a Warehouse and missing out on all that innovation. This is technical debt around future-proofing your business that you do not want to delay. It is also needed for increased agility to on-board new use cases and take them to production faster. Even if we stretch our imagination and assume that they could run ML, they would only be able to run it on structured or semi structured data which would be highly limiting. As we have noted in previous chapters, the bulk of the competitive hidden value is in the unstructured data that includes text, image, audio, video data that a Warehouse will not know what to do. Some may claim to handle it, but what they really do is keep the data in a cloud format and embed the link in the warehouse. That certainly does not qualify as support for unstructured data.

Now, let's layer in streaming, otherwise known as real-time use cases. This is the velocity aspect of big data systems, which were an after-thought for warehouses because streaming workloads were not mainstream in the 1980s when these systems were designed.

We have identified some of the shortcomings of warehouses. That does not mean that the data lake, which came later in the evolutionary trajectory, solved every problem. In the next section, we will look at which of these challenges it could address and where it fell short.

# Discovering when a data lake does not suffice

Data lakes were supposed to fix all the deficiencies of warehouses, but did they? Let's find out. The best contribution of data lakes was to truly unify all kinds of data in an open format at all velocities, and this includes real-time ingestion and analytics. Yes, a big check mark for this one. We can now have all kinds of unstructured data right alongside the structured ones, making them true first-class citizens. Likewise, we can enable ML on this data along with having the ability to use high-level languages to grapple with data using open APIs. Also, another big advantage is that data remained in an open format, allowing all tools of the ecosystem to leverage it effectively from its single source of truth. No more vendor lock-in!

The thing that they had to give up was effective BI, which warehouses were so good at. **Exploratory Data Analytics** (**EDA**), which did not have such stringent SLAs, was enabled, but core BI reporting and dashboarding was slow and did not scale for the hundreds of analysts wanting to tap into all the data in the lake. So, special curated extracts were created and pushed into BI tools, such as Tableau and Power BI, which of course compromised end-to-end latency and data freshness. Data science and ML received a lot of attention in the early 2000s, which is what kept data lakes around. It served the purpose of a lab where data scientists could bring in their new tools and experiment to their heart's content. A good discovery calls for a nice POC, or a research paper, but till this makes its way into production, the value will not really be realized by the business sponsoring this effort. However, concern around performance and inherent complexity could not be contained for too long. The data was there, but the analytic infrastructure on top of the consolidated data was too painful to use. People were not getting the perceived value and were growing frustrated with their investment.

Great, we've now enabled AI and some BI on all the data. So, why does this not suffice? It is because the rules were too loosely defined. Very often, the ingestion layer was simplified at the cost of consumption, so data lakes turned from silos to swamps. We've examined these fallouts in previous chapters, so we will not be revisiting them here. Suffice to say that data **reliability**, **quality**, and **governance** were overlooked, and Delta provided the foundational layer for addressing each of them. Security is applied at the file level, making it difficult to do fine-grained access at the row and column levels. You either have access to the entire file or you don't – there is no option to meet halfway.

Starting with support for ACID transactions, schema evolution, efficient indexing, compacting, file skipping, time travel, and a host of other features out of the box, Delta helped save investments made in data lakes and reaped value from the consolidated data and all the transformations made on them to get them to be analytics-ready. Delta is the glue between data lakes and the lakehouse architecture, truly delivering on the promise of a unified architecture for AI (ML, BI, and SQL) in batches or real time.

Apart from breaking the silos, there is the additional advantage of being able to link the two disparate forms of data through keys not truly classified as private and foreign key relation, because it just might not exist. However, other dimensions, such as geography, location, time, interests, and preferences, do help to connect the dots. Sometimes it is deterministic, and other times it is fuzzy, but the possibility has been opened up. Can you imagine examining cancer cell images in a data warehouse? Hell no! But with a data lake and its ability to soak up massive amounts of diverse data over long periods of time, the research community has greatly benefited – whether it is clinical trials or disease detection and prevention, or whether it is weather prediction or wildlife extinction prevention, or foot traffic to detect shopping patterns – yes, all of these disparate use cases can be tackled on a lake, something that was not previously conceivable in a warehouse.

It is no surprise that a lot of people will have a Data Warehouse on top of their Data Lake in their architecture to complement the deficiencies in each. Moving large volumes of data across disparate systems brings in fragility and compromises data quality.

# Addressing concurrency and latency requirements with Delta

Analytic queries are of two types:

- Ad-hoc data exploration by analysts as they proceed with data discovery activities. Data scientists and BI analysts have some tolerance for ad-hoc queries, meaning if it takes longer to retrieve the results, it is undesirable but tolerated.

- Known queries for well-defined consumption patterns. There is very little tolerance for known queries. Consumers expect these to be refreshed quickly as the end user may be a business executive or someone outside of the data organization who'll dislike the latency.

We should remember that a dashboard hosts several queries as widgets or sections and there are several consumers of that data. The time it takes to return the results is referred to as the **latency** of the query, and the maximum number of simultaneous users that it serves at the same point in time is referred to as the **concurrency** of the query. Latency and concurrency are the two KPI measures of BI queries and are used to compare and contrast different BI platforms. Moreover, latency and throughput (the volume of data processed in a given time unit) are two sides of a query. Because Spark was designed to handle large volumes of data, it was initially designed for throughput. On very large datasets, Spark wins hands down.

Traditional data lakes provide the necessary scalability, but not the real-time concurrency and latency needed for BI use cases. Delta comes to the rescue once again by providing performance at scale with a host of optimization techniques, such as caching, data compaction, and indexing. Previously, a subset of the curated data would be pushed to a warehouse to satisfy the latency and concurrency requirements of known queries. What this meant was that if a consumer needed a different access pattern or a slightly older dataset that was not available, they would have to request that their IT or data team get involved. This took data democratization a step backward. Ideally, we should allow people to access any data that they have privileges to. Delta Lake goes a step forward and allows BI tools to access data directly from the lake instead of accessing a sliver of the data in their expensive warehouses.

Delta was born in Databricks and given to the open source community. While benchmarks are to be taken with a pinch of salt, a recent benchmark that was done by the Barcelona Supercomputing Center shows Delta to be a winner in the 100 TB TPC-DS industry standard, which is widely regarded as the gold standard of performance benchmarks for comparing data warehouse performance. It should be noted that Databricks as a managed platform has a few secret sauces for higher performance compared to the open source Delta. However, the general trend of the numbers will be comparable.

Please refer to the following link to compare the query performance of Delta with other industry warehouses: `https://databricks.com/blog/2021/11/02/databricks-sets-official-data-warehousing-performance-record.html`.

# Visualizing data using BI reporting

One of the main use cases served by warehouses is around SQL queries and visualization in various reports and dashboards. Visualizing data involves converting it into graphical representations that make it easy to comprehend and detect outliers. It is especially useful to correlate data and make a statement about it that can be remembered for a long time. As noted earlier, the queries themselves fall into the two categories of *ad hoc data exploration* and *known queries*. There are a lot of tools on the market for visualizations, including Tableau, Power BI, Looker, Spotfire, and Qlik. They allow non-technical users to easily build personalized reports and dashboards, provided that there is access to the right datasets.

Spark supports a distributed SQL engine using its Thrift-based JDBC/ODBC or command-line interface (`https://spark.apache.org/docs/latest/sql-distributed-sql-engine.html#running-the-thrift-jdbcodbc-server`).

Managed platforms such as Databricks offer this as Databricks SQL capability catering to the analyst persona. Apart from the ability to write SQL queries against Spark clusters, there are other value-adding features of in-built visualizations, dashboards, alerts, query history, integration with a data catalog, and ACL-based permissions to data access.

Because a data warehouse can get very expensive, traditionally it retained a limited amount of data so that analysts could tap into small subsets of the original data. This is not conducive to long-term trend analysis. Moreover, BI tools accessing the warehouse are slow because data-fetching from a SQL endpoint is usually single-threaded. The resulting data is often packaged as an extract and stored directly on the BI tool for fast subsequent user access.

The inherent capabilities of Delta allow analysts to access all the data in a data lake. With Delta, you can now connect your preferred BI tool directly to your data lake and benefit from the better performance, leading to lower latency and improved concurrency. Existing authentication solutions can be leveraged to ensure that the right analyst has the right access privileges. Improved ODBC/JDBC drivers reduce the number of round trips needed for metadata access, and improved bandwidth and data transfer rates help push larger volumes of data in parallel, overcoming the single-thread bottleneck. Some vendors have built optimum connectors to squeeze as much data transfer out as possible to make the experience better. This is a list of Delta connectors available today, and the list keeps getting bigger: `https://delta.io/integrations`.

You can continue to use the languages, services, connectors, or databases of your choice. For example, the Power BI connector can read Delta tables natively using the PowerQuery/M function from any storage that it supports.

The following diagram shows the main components involved in the query orchestration in a managed platform like Databricks, starting with the BI users on their favorite BI tool issuing a query over a JDBC/ODBC driver to a compute endpoint where their query filter parameters get translated to predicate pushdown parameters. Compute is then used to pull the right data pieces, and at this point, the results can be passed back to the user. An alternate strategy – especially when a large amount of data is involved – is to create storage for the extract. Using Apache Arrow to serialize the output, this can be written efficiently into disk, and then the endpoint is excused from the retrieval path. The BI tool can directly access the storage layer and consume a large amount of data in parallel streams. The endpoint will just return the link to storage, greatly relieving the memory and disk-spill needs. To summarize, smaller datasets, say <1 MB, can be returned directly, but larger ones consisting of several megabytes or gigabytes can fall back to retrieval directly from storage. This ensures that smaller datasets' performance is not sacrificed in favor of that of larger ones.

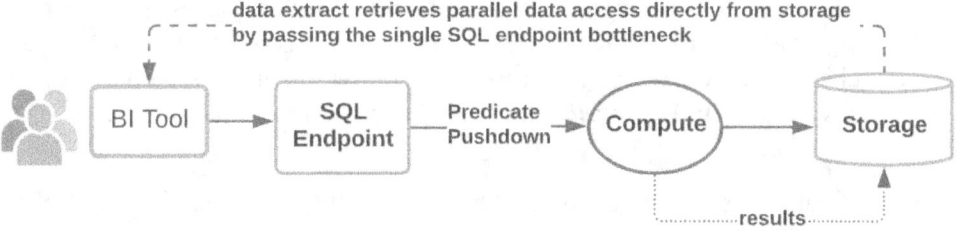

Figure 7.3 – How BI tools tap into all the data in your lake

Some of the common requests once these have been built relate to the ability to refresh data periodically, expose some parameters of the query, and the ability to share these dashboards and insights. These are offered out of the box in managed platforms or are built using the foundational blocks.

You are not confined to BI tools for visualizations. There are a host of libraries that offer rich capabilities. Your choices depend on the language you are using. In Python, the popular ones are Matplotlib, Seaborn, Bokeh, and Plotly; in JavaScript, D3 and Chart.js; in R Ggplot2, Lattice and Rgl are popular; in Scala, Vegas and Breeze-viz. When making a choice, you need to think about what kind of charts you want to build, how large your dataset is, how much customization is possible, what the learning curve is, and what your performance requirements are, especially on mobile.

## Can cubes be constructed with Delta?

Anybody who has worked with warehouses recognizes the widely used star schema and the snowflake schema. As a quick reminder of what they are, let's take the example of a retail store selling products. The main fact table is the transaction table being updated continuously. The dimension tables supporting it are details around Customer, Product Catalog, Price Points, and Date. In a star schema, the fact table is in the middle, surrounded by the dimension tables that it joins with. In a snowflake schema, instead of having a single layer of dimension tables, there may be multiple layers – that is, one dimension table will refer to other sub-dimensions. For example, the Customer dimension may further be broken to Address that it refers to extending the radius to another dimension. This is definitely more complex, and additional joins are expensive, especially with big data, which is why star schemas are more popular. This was introduced by Ralph Kimbal in the 1990s to help denormalize business data into dimensions and facts and avoid repetitive additions to data fields.

**OLAP** stands for **Online Analytical Processing** and is used by business analysts to perform dimensional analysis by aggregating, joining relevant data pieces, and doing intensive precomputation of results for fast consumption. This is typically referred to as *cubes*. OLAP cubes require data to be structured in a star or snowflake schema. They have some variations, such as *ROLAP for relational, MOLAP for multi-dimensional, HOLAP for hybrid*, and many others.

In the context of data warehouses, it is natural to ask how these cubes are built in Delta.

If Henry Ford were to ask his customers what they wanted, they would have said "faster horses," but he understood what they really wanted and invented the moving assembly line to build affordable cars that could take them from one place to another faster and more reliably than a horse could. The "what" may not change as often as the "how," since we have the luxury of being in an era of fast tech innovation, so when people say they want cubes, what they want is fast access to analytics. Cubes, like horses, may or may not be as relevant. With the new tools under our belt, let's see how we can serve the analytic needs. Sophisticated columnar technologies and fast access to data should allow us to meet in the middle, meaning there is some aggregation, but it may not be as complete as cubes because the size of data has exploded and it would not be wise to spend compute cycles on generating pre-aggregates that will never be used! This is referred to as the probability of access, and in the context of big data, this is an important consideration since design decisions involving tradeoffs have to be made on its basis. What is needed is access to data at any granularity, not just predefined or aggregated. The creation of star/snowflake schemas may not be needed. A simple change to a cube may warrant an entire re-computation, which is expensive, unnecessary, and certainly not sustainable.

Let's see how we can address the need with Delta. The creation of fact and dimension tables can proceed as usual with Delta tables. Z-order refers to multi-dimensional clustering and is the silver bullet in this case. Rows from the Z-order definition are collocated in as few files as possible. So, instead of scans, we now have an option to do seeks. Joins are to be carefully considered. Too-small or too-large files hinder performance when accessing data. If the file size is in the goldilocks zone, the dynamic file-pruning will skip to the right file or files. A good size range for files in our use case is 32-128 MB.

You can set the file size in Python or SQL using the following snippet:

```
#Python
spark.conf.set("spark.databricks.delta.optimize.maxFileSize",
<size in bytes>)
```

You can set the file size in SQL using the following snippet:

```
SET spark.spark.databricks.delta.optimize.maxFileSize=<size in
bytes>
```

You can set it only for a specific table using the following command:

```
ALTER TABLE <database>.<table> SET TBLPROPERTIES
(delta.targetFileSize=33554432)
```

If the tables have already been created, you can still set the `table` property, and the next `ZORDER` operation will take care of it.

If you have already added a `ZORDER` clause during table creation, you can still add and/or remove a column to force a re-write before arriving at the final `ZORDER` configuration.

Facts tables are humongous, containing petabytes of data. The careful choice of partitions and Z-order fields will ensure fast access. The `OPTIMIZE` command can be run periodically for efficient file compaction and pruning. You may be tempted to Z-order all fields, but that is a bad idea. Z-order is most effective for one to four fields, so some thought needs to go into choosing the right fields. The recommended approach to optimizing the fact table is as follows:

```
%sql
OPTIMIZE <FACT_Delta_TABLE>
ZORDER BY (LARGEST_DIM_FK, NEXT_LARGEST_DIM_FK, ...);

OPTIMIZE <DIMENSION_Delta_TABLE>
ZORDER BY (BIG_DIM_PK, LIKELY_FIELD_1, LIKELY_FIELD_2);
```

Big data usually does not enforce **primary key/foreign key** constructs, but having facts and dimensions means the existence of some surrogate keys that can be relied upon to be unique to help with the joins. If the dimension table is small, it can be broadcasted as-is to all the worker nodes. Since fact tables join against dimension tables, the likely candidates for Z-order in the fact table are the foreign keys, and the most likely fields that are fetched from the dimension tables should be Z-ordered as well.

At this point, we have gone over what data use cases a warehouse solves, some relevant scenarios around AI, unstructured data, and non-SQL workloads that are not conducive to warehouses. That said, data lakes pose some challenges as well. Some enterprises follow the middle path and have both as part of their architecture. In the next section, we will examine the recommended relative placements of these technologies should you choose to include both, and we will ponder ways to collapse them in order to simplify the architecture.

# Analyzing tradeoffs in a push versus pull data flow

A long, long time ago, we started with a data warehouse. As we discovered its inadequacies, we moved to a data lake. However, a vanilla data lake is no silver bullet, so folks would perform expensive ETL in a data lake and push curated, aggregated data slivers into a downstream warehouse for BI tools to pick up. Another architecture anti-pattern that we've seen in the field is ETL being done in a warehouse and pushing data to a lake to do ML. We have come a long way from there. Modern data lakes embrace the lakehouse paradigm, and BI tools can directly reach out to the data in a lake, bypassing the warehouse completely. We believe that this pattern will continue to gain traction in the industry. So, is the warehouse dead? Yes, in spirit, it is, but in practice, it'll take a few more years to phase out completely. So, when is it good to have any kind of specialized data stored to the right of a data lake? If it can be avoided, that would be great. However, there are some scenarios that would benefit from a serving layer. Let's examine a few situations to see where it makes sense, if at all.

Using Redis or a similar cache for quick access to data, there are hundreds of attributes that can be searched and the updates are localized. It means that not all the data gets updated all the time. Redis offers quick access to the consumers across several access patterns where the filters vary wildly and a Z-order on all these fields may not be practical.

OLTP systems generate transactional data that gets fed into an analytical OLAP system for analysis. Sometimes, the output of the OLAP system needs to be fed back to the original transaction system that it came from. Let's look at some scenarios:

1. A user makes a purchase in a store, or perhaps online.

2. The transaction is committed, and a record appears in the OLTP system.

3. This makes its way onto an OLAP system either in a batch or real time.

4. The OLAP system ETLs the data and runs aggregations to roll the data along predefined dimensions to provide the marketing and sales department with the KPIs they need to determine the health of the business and make changes to strategy and process.

5. The inventory count, past sales, reviews, and recommendations are being tracked, and this information needs to be passed back to the OLTP system to be consumed by a different audience – the end user themself, who may use these analytics to make a decision on their next purchase.

The following diagram shows the different data personas interacting with the key data stores, each of which has its own characteristics and strengths for supporting a specific consumption pattern.

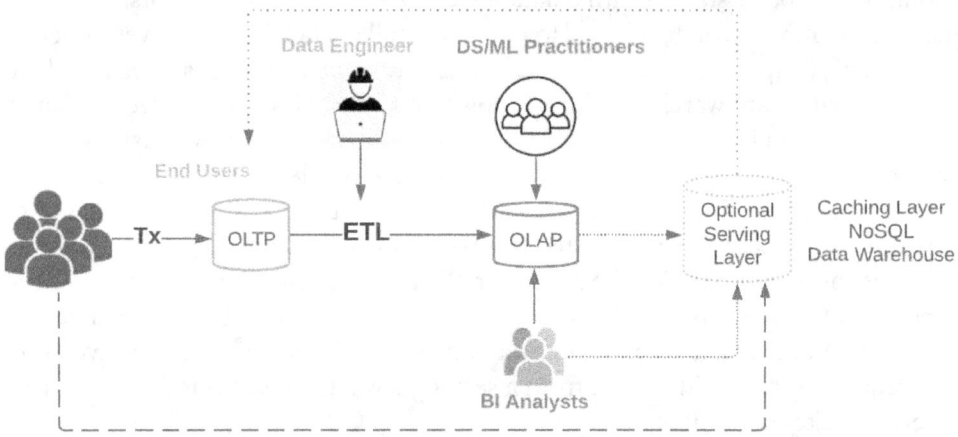

Figure 7.4 – Rendering data insights to different audiences for different consumption needs

The OLTP and OLAP systems are well established and well demarcated. The serving layer offers many choices and the decision is usually based on SLA requirements as there are always cost and latency tradeoffs to be considered. The serving layer is a good place to introduce a semantic layer that offers self-service capabilities along with large curated datasets for folks in the organization to leverage.

For example, Power BI Premium offers the ability to host 10-400 GB of compressed data in-memory and scale to support thousands of users, and it can be shared with the enterprise using Azure Active Directory security constructs. It favors star schema and joins on surrogate keys, but it can model other techniques such as flat files. Incremental refresh enables incremental data loading into import tables using managed date ranges and helps with data refresh. In the next section, we will examine how important it is for data to be in an open format.

# Why is being open such a big deal?

Today, every warehouse is trying to be a little more like a lake, and every data lake is trying to be a little more like a warehouse. It is hard to say which path is easier, but if one were to hazard a guess, then lake-to-warehouse probably wins since lakes offer much more in terms of capabilities. However, the bigger question is: is this a futile pursuit? Each has something good to offer, so the time may be right for the creation of a new category. In the next section, we will explore this in more detail, but first, let us perform one litmus test to distinguish a warehouse from a lake.

One characteristic that clearly delineates warehouses from lakes is their support for open formats and open APIs. All warehouses are prescribed to a proprietary format. So, why is **open** such a big deal? When everyone has access to your code, your developers can extend beyond the company walls to the entire developer community. Better ideas, more bug fixes, innovation, adoption, and higher growth and creativity can be crowdsourced. This also trains the community on your offerings. All of a sudden, every tool in the ecosystem can leverage the data in place. Data virtualization can be achieved with federated engines that pull data from different sources without having to ingest it, and having data in an open format helps with these federation efforts. Delta has its counterparts as well – Hudi from Uber and Iceberg from Netflix all vie for the same audience. They build their secret-sauce data engine and access layers on top of data in object storage, allowing for a level of flexibility and choice that is impossible when it comes to warehouses.

Being open allows for a plethora of APIs written in languages beyond SQL to access the data. Research requires well-curated datasets that are made available to people worldwide who use ML APIs to access this data. Imagine each of them having to pull this proprietary data and make a local copy to do some useful ML on top of it. An application built on top of these open APIs can claim to be truly portable across execution environments. Data sharing will be covered in a future chapter, but no company wants to reprocess openly available raw data over and over. Once curated, that data can be shared and monetized in data exchanges securely. If data was not in an open format, consumers would be restricted to only those on the platform. On the World Wide Web, that would be a narcissistic outlook. A company's partners and vendors should be allowed to consume data in an open format that is also efficient. Text files are a thing of the past – in big data, sharing data as CSV files will just not fly. Any conversion of large volumes of data requiring copies is a wasteful operation, in terms of compute, storage resources, and time. Delta allows for sharing data without the need to make a copy and without the other party needing to be on the same type of platform. This is only possible because Delta is in an open protocol!

# Considerations around data governance

Data governance refers to aligning all aspects of data strategy, business strategy, and compliance requirements. A three-pronged approach of people, policy, and process will provide oversight for all data operations from the time data touches a system to the point it leaves. Roles and responsibilities dictate who has access to what data, something that needs to be enforced and monitored. Data lineage is tracked to provide accountability for how data has been transformed at various steps. Delta's history functionality provides a good audit trail. A central catalog builds on top of it and provides a central place for defining the rules, enforcing them, and monitoring compliance via audit logs. Some of these catalogs have to be built and stitched together unless a managed platform that has taken care of these aspects is leveraged.

People using data need to be assured of its quality, so being able to define constraints, note when they have been violated and what percentage of the data is affected, and what to do if a violation happens are all aspects of data governance that a consumer of data must be informed of. Databricks has a feature called **Delta Live Tables** (**DLT**), which gives a declarative way of specifying the **Directed Acyclic Graphs** (**DAGs**) of dependent transformations and constraints on each intermediate data, along with giving guidance on what to do, such as retain the data with a reduced-quality metric, drop it, or quarantine it.

In big data systems, the metadata is big data as well. In the case of Delta, most of the metadata sits alongside the data in cloud storage.

# The rise of the lakehouse category

Simply put, "lakehouse" refers to an open data architecture that combines the best of data lakes and data warehouses on a single platform. At this point, it would be fair to say that a lakehouse is closer to a data lake than a data warehouse. In fact, it is an extension of your data lake to support all use cases, from BI to AI. All data science and ML personas who were shunted into downstream applications because the tools of their trade were so vastly different and can now share the same stage and have access to the same data as other data personas. This eliminates the need to stitch fragile systems together and leads to better data quality and end-to-end latencies since there is no need to copy data across disparate architectures. The following diagram shows the growing pains of both warehouses and lakes, and how a lakehouse is a combination of the best attributes of both architectures.

Figure 7.5 – From the tug of war between warehouses and lakes, the lakehouse emerges!

Let's compare the capabilities of each of these architectures across different dimensions:

- **Storage**:

  - **Data lake**: Provides open-format, but potentially lower-quality, data, leading to data swamps and coarser file-level access controls.

  - **Data warehouse**: Even though data is more reliable with fine-grained row-/column-level table access controls, the favored data type is structured data, and it remains in a closed, proprietary format.

- **Compute**:

  - **Data lake**: Although it is more economical, especially for large datasets, it comes with the price of higher operational complexity.

  - **Data warehouse**: Although easy to use and of high concurrency, scaling can get exponentially more expensive, resulting in dropping historical datasets to contain costs on limited approved storage.

- **Consumption** :

  - **Data lake**: Open, with a rich ecosystem of languages, tools, and frameworks. The most prevalent BI use cases are not considered first-class citizens when it comes to a data lake.

  - **Data warehouse**: Although SQL-based BI use cases are favored, there is little to no support for data science and ML use cases.

Every couple of years, the IT department of every business takes a hard look at its infrastructure and assesses its current technical debt to make the next investment. Complex, outdated, and inadequate infrastructures are constantly being replaced by leaner, meaner architectures. Forty years ago, it was warehouse technology, 10 years ago it was the data lake wave with Hadoop leading the charge, and now we're in the era of the lakehouse. Databricks is credited with ushering in this new category, and it is being embraced by all of the cloud vendors. Bill Inmon, father of data warehousing, wrote in his book *Building the Data Lakehouse* about this evolution and why it is time for a simpler approach to future-proofing your business by embracing the lakehouse paradigm. No one has a crystal ball they can use to predict what use cases will be needed tomorrow, so future-proofing your underlying architecture will brace your organization to handle any use case in the fastest amount of time, sticking to open standards and open formats so that subsequent generations will not be burdened by any technical debt accrued from inadvertent bad choices. We started this chapter with a quote from Darwin, reminding us to be flexible and adaptable, and that is exactly what the lakehouse architecture does, which is why we believe that it'll stand the test of time.

The following diagram is a reminder that data from disparate sources can come together and make the whole greater than the sum of its parts. The real value is reaped as data moves along the pipeline from bronze to silver to gold and gets blended with other data sources to give insights, which would not be possible if we treated each pipeline in isolation.

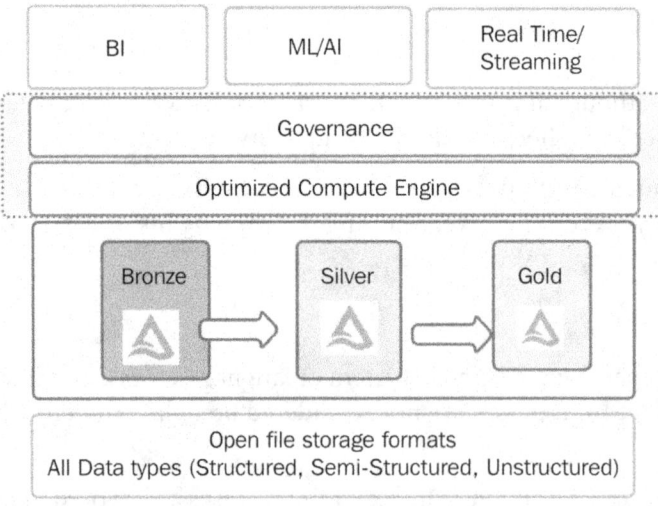

Figure 7.6 – The lakehouse stack

Let's examine the components of the lakehouse stack:

- The bottom-most layer of the stack is raw object storage.

- The layer right above is the metadata transactional layer, which provides more usable value to the data.

- The next two layers above the metadata transactional layer can either be built in-house, in the open source world, or a managed service offering can be leveraged to provide that functionality in a typical build-versus-buy situation. This typically includes an optimized compute engine that is well governed by the defined rules, so as to comply with business needs.

- That leaves us with the top-most layer. This enables all types of use cases to be built on top of it, thereby taking care of existing scenarios while future-proofing the ones left unidentified so far.

Let's now summarize the key differences between lakehouse and warehouse architectures along several dimensions of data metrics that we learned in this chapter.

# Summary

In this chapter, we emphasized the need to choose the right architecture for future-proofing a business. This choice will determine the future agility of on-boarding use cases and the productivity of data personas in exploring and executing use cases. Traditional data warehouses and data lakes have their own strengths and weaknesses, and the lakehouse is a happy amalgamation of the two technologies.

The data format of warehouses is closed and proprietary, whereas a lakehouse prescribes an open data format and the recommendation is to use Delta, as it is the best open source data format in the open source community today. The data type of warehouses caters to mostly structured data, and some semi-structured, whereas a lakehouse supports all kinds of data, including unstructured. Cloud storage is highly scalable, durable, and cost-effective, so a lakehouse is not only highly scalable but much cheaper and more performant than its warehouse counterpart. A warehouse was designed for BI/SQL workloads, and that is the only use case it handles well, whereas a lakehouse handles the whole breadth of use cases spanning BI, ML, real time, and SQL to provide a unified platform and a single view of all the data. There is no longer a need to think about which copy of the data in which system is the right version to use. The open format enables open APIs for direct access to data files using multiple languages such as Python, R, Java, Scala, and SQL. The other two things that warehouses were good at, which data lakes took time to adopt but lakehouses provided for from the get-go, relate to high-quality, reliable data with ACID transaction support and fine-grained security and governance for row- and column-level table access.

The key ingredient in this is Delta, an open source project that delivers the missing ingredients of your data lake, namely, reliability, quality, performance, and governance on cloud storage. There is rapid adoption among various industry verticals because it is completely agnostic to specific use cases and offers something for everyone.

In previous chapters, we focused on the data engineer persona building data pipelines. In this chapter, we focused primarily on the data analyst persona using SQL to explore data and build reporting dashboards. In the next chapter, we will look at some atypical data scenarios that needed special handling, and in *Chapter 9, Delta for Reproducible Machine Learning Pipelines,*we will talk through the unique challenges that ML practitioners face and how Delta aids them at every stage of building an ML pipeline.

# 8
# Handling Atypical Data Scenarios with Delta

*"Every problem has a solution. Sometimes it just takes a long time to find the solution – even if it's right in front of your nose."*

*– Daniel Handler, American author and musician*

In the previous chapters, we established the need for a Lakehouse architecture paradigm to handle a wide range of use cases, from BI to AI. Data wrangling by itself may not be sufficient to get the data ready for consumption. Several conditions need to be addressed to ensure not only that the data is cleansed and transformed as per the business requirements but also that it is fit for the use case at hand. So, even when the logic of the pipelines has been ironed out, other statistical attributes of the data need to be addressed. This helps ensure that the data patterns for which it was initially designed still hold and are making the most of the distributed compute that it runs on.

In this chapter, we will explore cases beyond the happy path to see how Delta makes them simpler. For example, you will learn how to defend the insights that are generated when dealing with class imbalance, how to effectively use compute resources when dealing with skews in the incoming data, and how to balance the tradeoffs between bias and variance and unknown data patterns. You will learn when it is OK to evolve the schema and when a red alert should be raised to review the data and make adjustments to the pipeline, perhaps because the norm has changed and so the baseline assumptions and benchmarks need to change as well.

In this chapter, we will cover the following topics:

- Emphasizing the importance of **exploratory data analysis (EDA)**
- Applying sampling techniques to address class imbalance
- Addressing data skew
- Providing data anonymity
- Handling bias and variance in data
- Compensating for missing and out-of-range data
- Monitoring data drift

Let's look at some of these real-world data concerns, including how to detect them and address them effectively.

# Technical requirements

To complete this chapter, make sure you have the code and instructions that are located in this book's GitHub repository: `https://github.com/PacktPublishing/Simplifying-Data-Engineering-and-Analytics-with-Delta/tree/main/Chapter08`.

The examples in this book cover some Databricks-specific features to provide a complete view of the available capabilities. Newer features continue to be ported from Databricks to the open source Delta.

Let's get started!

# Emphasizing the importance of exploratory data analysis (EDA)

Data quality problems cost US businesses more than $3 trillion a year (reference: `https://hbr.org/2016/09/bad-data-costs-the-u-s-3-trillion-per-year`). In the previous chapter, we examined the capabilities of Delta, such as ACID transactions and schema evolution, which help ensure a high degree of data integrity as data is being processed. But what about the characteristics and temperament of the raw data itself? If it is riddled with holes and gaps, then using it to build a model will result in suboptimal, if not inaccurate, insights. Understanding the quality and reliability of the working datasets is an important step and should not be skipped.

EDA refers to the process of statistical analysis to review the source data and understand its structure, content, and interrelationships to help identify the true potential for data projects. This is where profiling the data is important as it produces critical insights.

`df.describe()` and `df.summary()` are APIs that you can call on the delta DataFrame to give you a bird's-eye view of key statistical metrics around your data columns:

```
1   from pyspark.sql.functions import *
2   df.describe('gender', 'salary').show()
```

▶ (2) Spark Jobs

▶ 🖿 df: pyspark.sql.dataframe.DataFrame = [id: integer, firstName: string ... 6 more fields]

```
+-------+--------+-------------+
|summary|  gender|       salary|
+-------+--------+-------------+
|  count|10000000|     10000000|
|   mean|    null| 72633.0076033|
| stddev|    null| 20003.2293585|
|    min|       F|       -26884|
|    max|       M|       180841|
+-------+--------+-------------+
```

Figure 8.1 – Descriptive statistics on the DataFrame

People who are familiar with the pandas DataFrame APIs will find a lot of similarities in terms of functions with Spark and sometimes prefer to convert between the two DataFrames using Apache Arrow. In the next few sections, we will examine the various ways to profile data to understand its inherent quality.

# From big data to good data

Andrew Ng, founder and CEO of *Landing AI* and former head of Google Brain, talks about the importance of data-centric AI over a model-centric AI in the pursuit of improving performance and quality of insights. He compares the process of creating an analytic asset to that of preparing food, where 80% of the time is spent in the prep phase and 20% of the time is spent cooking. This is true of training a model – the bulk of the time is spent ensuring the data is of high quality. He argues that data is food for AI and that after a certain baseline level, improving the data is more beneficial than improving the algorithmic code (reference: `https://www.youtube.com/watch?v=CFEJkVuHhRM`).

Let's see what is meant by *good data*.

Good data is defined as follows:

- It has unambiguous labels.
- It provides good cover for important use cases.
- It gathers timely feedback from production data (distribution covers both data drift and concept drift).
- It is sized appropriately.
- Good data is self-describing with meaningful comments in the metadata

Now, let's look at different ways of examining data to understand its quality.

# Data profiling

**Data profiling** refers to the process of examining, analyzing, and creating useful summaries of data. It yields a high-level overview that aids in the discovery of data quality issues, risks, and overall trends. Some of the advantages of data profiling are as follows:

- Better data quality and credibility
- Predictive decision making
- Proactive crisis management
- Organized sorting of priorities

Some common approaches are as follows:

- Structure discovery:

  - This helps to determine whether data is consistent and formatted correctly.

- It uses basic statistics to provide information about the validity of data.

- Content discovery:

  - This focuses on data quality. Data needs to be formatted, standardized, and properly integrated with existing data in a timely and efficient manner.

- Relationship discovery:

  - This identifies connections between different datasets (that is, join criteria).

EDA can be done using pandas profiling on pandas Data Frames for smaller data sets that fit on a single node. `spark-df-profiling` can be used on larger datasets for Spark DataFrames (reference: `https://github.com/julioasotodv/spark-df-profiling`). For each column, several statistics are generated in an HTML report. The following is an example of the output:

```
import spark_df_profiling
df = sqlContext.read.format("delta").("<path>").cache()
df.describe().show()
report = spark_df_profiling.ProfileReport(df)
```

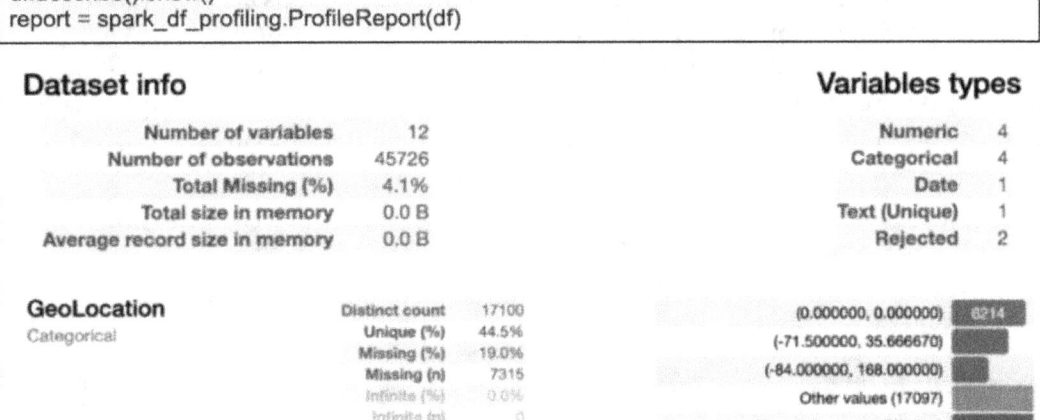

Figure 8.2 – Data profiling

For each column, several statistics are generated. Some are bare essentials such as the data's type or unique and missing values; quantile statistics such as min, max, range, and interquartile range; descriptive statistics such as mean, median, mode, standard deviation, sum, and skewness; and the most frequent values and histograms.

# Statistical analysis

Several techniques are used to study data for statistical soundness, including factor analysis, cluster analysis, variance analysis, principal component analysis, and many others. At a high level, they are divided into two broad categories – univariate and multivariate analysis.

- **Univariate analysis**:

  - Univariate involves analyzing a single variable in isolation.

  - This is where each column is examined to see how populated or sparse it is, what the range of values is, whether they are coming from different sources, and whether they are normalized and standardized.

- **Multivariate analysis**:

  - Multivariate analysis examines two or more variables at a time. Most multivariate analysis involves a dependent variable and multiple independent variables.

Spark provides several top-level APIs to understand the relationships between various columns. **Covariance (cov)** and **correlation (corr)** are related statistical terms. While covariance is a measure of how two variables change because of each other, correlation is a normalized measure of covariance that is easier to understand as it provides quantitative measurements of the statistical dependence between two random variables:

```
1   print(df.stat.cov('id', 'salary'))
2   print(df.stat.corr('id', 'salary'))
```

▸ (2) Spark Jobs

6737933.84222936
0.00011668559071227281

The statistical functions can be accessed directly from the SQL library or MLlib, as shown here:

```
1   from pyspark.mllib.stat import Statistics
2   import numpy as np
3
4   def parse_data(record):
5       return np.array([float(x) for x in record])
6
7   vector_data = df.select('id', 'salary').rdd.map(parse_data)
8   summary = Statistics.colStats(vector_data)
9   print(summary.mean(), summary.variance(), summary.numNonzeros())
10
11  print(Statistics.corr(vector_data, method='pearson'))
```

▸ (5) Spark Jobs

```
[5000000.5          72633.0076033] [8.33333417e+12 4.00129185e+08] [10000000. 10000000.]
[[1.00000000e+00 1.16685591e-04]
 [1.16685591e-04 1.00000000e+00]]
```

(Reference: `https://spark.apache.org/docs/latest/mllib-statistics.html`)

Although most real-world research examines the impact of multiple independent variables on a dependent variable, many multivariate techniques, such as linear regression, can be used in a univariate manner to examine the effect of a single independent variable on a dependent variable. Displaying the results in an understandable format is especially important in multivariate analysis because of the greater complexity of these techniques. Let's look at some popular visualization analysis preferences for both:

- Univariate:

  - **Bar Charts**: These are useful for making comparisons between categories of data or different groups of data. It helps to track changes over time.

  - **Pie Charts**: These provide an overview of a group of data by breaking it into smaller pieces and showing them in a different slice of the pie.

  - **Histograms**: These are similar to bar charts, except that histograms focus on the distribution of variables that have been binned instead of making pure comparisons.

  - **Frequency** Distribution Tables: The frequency distribution reflects the frequency of an occurrence in the data.

- Multivariate:

  - **Scatter plots**: These show the measure of the influence of one variable on another.

  - **Regression analysis**: This is used to analyze how two or more variables are related to each other.

  - **Correlation coefficients**: These analyze if the variables are related. "0" suggests that the variables are not related to each other, while "+/-1" reveals a positive or a negative correlation, respectively.

Now that we've established the baselines, let's learn how to compensate for certain scenarios, such as class imbalance, bias, variance, and data skew, among others.

# Applying sampling techniques to address class imbalance

Let's look at a scenario where there's a cell culture dataset that is being analyzed using **machine learning** (ML) algorithms to predict the onset of cancer. Most cells are normal; a small percentage may be abnormal. The two primary classes here are "normal" and "abnormal." This is an imbalanced dataset. This applies to multi-class datasets as well. An imbalance occurs when one or more classes have low proportions in the training data compared to other classes. Since the ML process involves "learning" from the dataset, there is a lot to learn about the normal scenarios and very little about the cancer ones. Most ML algorithms for classification are designed and demonstrated on problems that assume an equal distribution of classes and are designed to maximize accuracy and reduce error. The consequence of this imbalanced dataset is that the model is biased. Sometimes, it goes undetected and a model is built. Since the data was not sound, there will be telltale signs of imbalance in the model as well.

Imbalance introduces majority and minority classes. Sometimes, the imbalance is modest, while in others, it is severe. Classification problems that can have a severe imbalance in the class distribution across industry verticals include fraud, claim, anomalies, and intrusion detection. For example, less than .1% of credit card transactions are fraudulent. Understanding the latent behavior of the attacker and their pattern over long periods is a challenge. The same is true in the case of claims in the insurance sector, anomaly detection in a manufacturing process, or intrusion detection in a cybersecurity use case. Let's learn how to detect an imbalanced dataset.

# How to detect and address imbalance

The simplest way to check for imbalance is to count each of the dependent categorical values. A wide variation indicates imbalance:

```
1   from pyspark.sql.functions import countDistinct
2   countDistinctDF = (df.select('gender', 'salary')
3                              .groupBy('gender')
4                              .agg(countDistinct('salary').alias('class')) )
5   display(countDistinctDF)
```

▶ (3) Spark Jobs

▶ ▦ countDistinctDF: pyspark.sql.dataframe.DataFrame = [gender: string, class: long]

**Table**     Data Profile

|   | gender | class |
|---|--------|-------|
| 1 | F | 128370 |
| 2 | M | 127369 |

In the case of classification problems, we can examine the confusion matrix to check if any element is zero. As a reminder, a confusion matrix is a visual representation of an error matrix of predicted and actual outcomes. Now that we have analyzed the data and detected an imbalance, how do we deal with it?

- Augmenting the performance metrics beyond just accuracy will give a better picture. A good one to start with is the confusion matrix – don't study just the correct results but the incorrect ones as well. Precision, recall, and the F1 score are all good metrics to consider. These can be defined as follows:

  - **Precision (positive predictive value)** is computed by $TP/(TP+FP)$, where $TP$ stands for **true positives** and $FP$ stands for **false positives**.

  - **Recall (true positive rate)** signifies completeness/sensitivity and is computed by $TP/P = TP/(TP+FN)$, where $P$ stands for all **positives**, $FN$ stands for **false negatives**, and $TP$ stands for **true positives**.

  - **F1 score** is a weighted average of $P/R$, where $P$ is **precision** and $R$ is **recall**.

- Sometimes, using a different algorithm helps. For example, tree-based algorithms help in imbalance scenarios.

- In most cases, resampling strategies or data generation techniques such as **Synthetic Minority Oversampling Technique** (**SMOTE**) or **Adaptive Synthetic** (**ADASYN**) are used. Oversampling refers to taking multiple copies, while undersampling refers to dropping data points. Oversampling the minority class and undersampling the majority class are both viable options. The other approach is to generate synthetic samples. You should always split the data into test and train sets before trying oversampling techniques!

In the next section, we will look at synthetic data generation techniques in more depth.

## Synthetic data generation to deal with data imbalance

Synthetic data generation not only helps address data imbalance scenarios but also provides test data in cases where real production data is not sanctioned to be used in non-production environments. Spark provides several APIs to generate random data. An example of generating uniform and normal distribution data in a new column is as follows:

```
1  from pyspark.sql.functions import rand, randn
2  df.select("id", rand(seed=10).alias("uniform"), randn(seed=27).alias("normal")).show()
```

▶ (1) Spark Jobs

```
+-------+-------------------+--------------------+
|     id|            uniform|              normal|
+-------+-------------------+--------------------+
|6280266| 0.1709497137955568|  -0.8664700627108758|
|6280267| 0.8051143958005459|  -0.5970491018333267|
|6280268| 0.5775925576589018|  0.18267161219540898|
```

If we are going to generate new data points, then it cannot be done at random. It should mimic the behavior of the real world. There are libraries such as Faker (reference: `https://faker.readthedocs.io/en/master/fakerclass.html`) that can be used to generate data quickly and they can be provided with a range of values, but this is not realistic. For example, one line item could be a street address and a ZIP code, and a data point may associate the same address with a different ZIP code. So, how can we simulate real-world data without creating multiple copies – that is, oversampling? The answer lies in understanding latent inherent data characteristics and profiling them first. The SMOTE technique finds the n-nearest neighbors in the minority class for each of the samples. Then, it draws a line between the neighbors and generates random points on the lines. These data points are now more realistic than those generated purely at random.

This can be a time-consuming and expensive process. The Data Generator is a nifty tool that uses a Python framework packaged as a wheel file from the public Databricks Labs Repo to help you. A schema and a seeding approach for each column must be specified as an initial spec that can be further tuned and altered. The "build" method is called on the spec to generate the data, which can then be saved in Delta tables. The following snippet shows how to use it:

```
1   import dbldatagen as dg
2   from pyspark.sql.types import IntegerType, FloatType, StringType
3
4   deltaDataPath='/tmp/chapter8/csvToDelta'
5   df_spec = (dg.DataGenerator(spark, name="test_data_set1", rows=1000000, partitions=4)
6                   .withIdOutput()
7                   .withColumn("r", FloatType(), expr="floor(rand() * 350) * (86400 + 3600)", numColumns=10)
8                   .withColumn("code1", IntegerType(), minValue=100, maxValue=200)
9                   .withColumn("code2", IntegerType(), minValue=0, maxValue=10)
10                  .withColumn("code3", StringType(), values=['a', 'b', 'c'])
11                  .withColumn("code4", StringType(), values=['a', 'b', 'c'], random=True)
12                  .withColumn("code5", StringType(), values=['a', 'b', 'c'], random=True, weights=[9, 1, 1]))
13
14  syntheticData_df = df_spec.build()
15  syntheticData_df.write.format('delta').mode('overwrite').save(deltaDataPath)
16  print(syntheticData_df.count())
```

▸ (6) Spark Jobs

▸ 🔲 syntheticData_df: pyspark.sql.dataframe.DataFrame = [id: long, r_0: float ... 14 more fields]

1000000

(Reference: `https://github.com/databrickslabs/dbldatagen/tree/v0.2.0-rc0-master`.)

In the next section, we will look at cases of uneven data distribution and its effect on system performance, as well as how these bottlenecks can be addressed.

# Addressing data skew

In Spark, data resides in different "partitions" that guide the decision of how to divide the data among different worker nodes to get the benefits of parallelism. In an ideal case, data in each of the partitions is divided equally so that the load on the workers is uniform and the cluster resources are utilized more efficiently. Data skew is a condition in which a table's data is unevenly distributed among partitions in the cluster. This has several negative consequences, namely a reduction in the performance of queries, especially those that involve joins. Joins typically result in shuffle and data skew, which can lead to a labor imbalance among the workers. This means that only a few workers are doing the heavy lifting, prolonging the query response time and resulting in unnecessary compute wastage. Let's look at the four main types of joins:

- **Broadcast Hash Join**:

  - Requires one side to be small.

  - No shuffle nor sort is involved, hence it is very fast.

  - `SELECT /*+ BROADCAST(a) */ id` FROM a JOIN b ON a.key = b.key.

- **Shuffle Hash Join**:

  - Needs to shuffle data but not sort it.

  - Can handle large tables, but will perform **Out Of Memory (OOM)** too if data is skewed.

  - One side is smaller (3x or more) and a partition of it can fit in memory (enabled by `spark.sql.join.preferSortMergeJoin = false`).

  - `SELECT /*+ SHUFFLE_HASH(a, b) */ id` FROM a JOIN b ON a.key = b.key.

- **Sort-Merge Join**:

  - This is the most robust type of join and it can handle any data size.

  - It needs to shuffle and sort data, so it is slower in most cases.

  - `SELECT /*+ MERGE(a, b) */ id FROM a JOIN b ON a.key = b.key.`

- **Shuffle Nested Loop Join (Cartesian)**:

  - Does not require join keys as it is a Cartesian product of the tables.

  - Avoid doing this if you can as it is the most expensive join type.

  - `SELECT /*+ SHUFFLE_REPLICATE_NL(a, b) */ id FROM a JOIN b.`

So, how do we determine if we have a skew problem?

If a query appears to be stuck finishing very few tasks (for example, the last two tasks out of 100), then that is an indication of data skew. Another place to check is in the task summary statistics. If there is a wide range between the 25th percentile and the 75th percentile, then that also points to data skew:

- Click the stage that is stuck and verify that it is doing a join.
- After the query finishes, find the stage that does the joins and check the task duration distribution.
- Sort the tasks by decreasing duration and check the first few tasks.

Now that we understand data skew, how can we address it?

Delta accepts skew hints in joins. A skew hint must contain the name of the relation with skew. A relation can be a table, view, or subquery. All joins with this relation then use skew join optimization.

Spark 1.0 was rule-based, Spark 2.0 added a cost-based optimizer, and Spark 3.0 optimized this further by adding the **adaptive query execution** (**AQE**) feature, which re-optimizes queries based on the most up-to-date runtime statistics, thereby taking care of handling data skew using three approaches. These are as follows:

- Dynamically switching join strategies:

  - Converts a sort-merge join into a broadcast hash join

- Dynamically coalescing shuffle partitions:

  - Shrinks the number of reducers and coalesces shuffle partitions

- Dynamically optimizing skew joins:

  - Dynamically handles skew in a sort-merge join and shuffles the hash join by splitting (and replicating it if needed) skewed tasks into roughly evenly sized tasks.

  - If the job encounters spills (memory to disk operations are expensive), increase the number of partitions using `spark.sql.shuffle.partitions`).

Adaptive query execution extends these optimizations beyond rule-based and cost-based optimizations by giving more decision-making power to the runtime, which can make decisions based on the statistics that are collected as the query runs. AQE can be enabled using `spark.sql.adaptive.enabled` alongside skew join (`spark.sql.adaptive.skewJoin.enabled`). The idea with AQE is that the query metrics are utilized in flight to tweak the execution plan dynamically. It eliminates the need for hints to influence the query plan. It is especially useful if statically derived metrics are inaccurate.

# Providing data anonymity

Several scenarios require sensitive user data to be protected and avoid exposure to entities that do not have the right entitlements. It could be **Payment Card Industry** (**PCI**) data of credit cards or **Protected Health Information** (**PHI**) data from health records, which, in the wrong hands, causes financial and reputational damage. Several data privacy techniques can be used to protect **Personally Identifiable Information** (**PII**) – some trivial and others more complex. The strategy should be carefully considered when you're dealing with large data as there is a performance price to be paid for the additional processing. Other considerations include the need for re-identification, read/write efficiency, the schema, and data format choices. Let's look at some of the strategies that can be used:

- Encrypting data at rest and in motion

- Hashing (for example, using `sha512()` for ultra-sensitive data such as passwords)

- Tokenization (in the form of mapping – for example, one-hot encoding)

- Row/column-level security (homegrown or enforced by tools such as Immuta, Privacera, Ranger, and others)

The terms **anonymization** and **pseudo-anonymization** are closely related and have a similar intent – that is, to protect data – but they aren't the same. Anonymization is a form of masking where the original data is irreversibly altered, whereas pseudo-anonymization replaces the original data with a pseudonym for re-identification later. In other words, there is a mapping between the original data and the pseudonym. The latter is used in the hands of unauthorized users, while the former is used for user-specific analysis where data needs to be joined correctly in the context of the actual entity. By using a combination of encryption libraries and **user-defined functions** (**UDFs**), PII data can be protected.

Sometimes, data is binned or generalized to retain some breadcrumbs to aid in analysis but still make it untraceable to the original data. An example of this is to retain the last four digits of a credit card or only the ZIP field of an address. The following table shows an example of using masking and scrambling to obfuscate the sensitive `host` field:

| **Masking** makes the data completely opaque | **Scrambling** retains some characteristics |
|---|---|
| val mask = (text: String) => {<br>    "*** Masked ***"<br>}<br>spark.udf.register("**mask**", mask) | val scramble = (text: String) => {<br>    if (text != null) {<br>        val lst = text.toList;<br>        val rand = scala.util.Random.shuffle(lst);<br>        rand.mkString("");<br>    } else<br>        "",<br>}<br>spark.udf.register("**scramble**", scramble) |
| Example of calling mask and scramble on the same Host field | |
| SELECT **mask**(Host) as maskedHost, **scramble**(Host) as Host FROM  algo.Out LIMIT 5;<br><br>maskedHost   ⬥   Host<br><br>*** Masked ***       nsee16-9.c.11o-ean7a-4427aad.pndyr810t64f4ebc54lae:ha-29rutriae1nsiak4eg-le.5mt-c | |

Figure 8.3 – Data anonymization

Sometimes, multiple copies of the data are made – one for the entitled user with all the fields and another for the unauthorized ones with the sensitive rows and columns suppressed. This is not sustainable because there may be several groups of consumers with different entitlement requirements. How many copies are we going to maintain and how do we reconcile them? Another strategy is to create different views on the same data. This is a little better than managing multiple copies but still hard to manage.

Managed platforms such as Databricks provide this functionality out of the box – it includes two user functions called `current_user()` and `is_member()` to express row and column-level access dynamically in the view definition, without the need to maintain multiple views for different users with different entitlements.

Let's look at an example of sales data where the raw delta table is `sales_raw`. We create a view called `sales_redacted` off it where only auditors can view the user's email and for others that value is redacted. End users are given privileges to access the data through this view. The following diagram shows two paths. The first path uses the UDF approach to obfuscate data, while the other path stores the data as-is in a bronze delta table where users do not have direct access. Instead, dynamic views are built on top to protect sensitive rows and columns. Additional data masking can be achieved using regex to identify standard sensitive data patterns. For example, an email will always have an @ symbol in it:

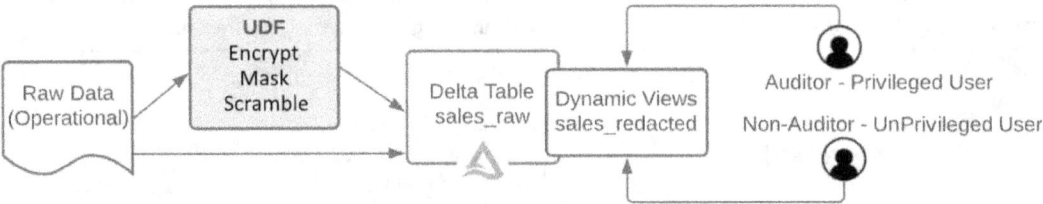

Figure 8.4 – Privileged and unprivileged data access

The code for creating the dynamic views in all three cases is as follows:

| Column Level Access | Row Level Access | Masking |
|---|---|---|
| Only auditors access email | Only managers can view rows with sales over a million | Regex obfuscates emails, allowing Analysts to see only the domain part of email |
| <pre>CREATE VIEW<br>sales_redacted AS<br>SELECT<br>  user_id,<br>  CASE WHEN<br>    is_member('auditors')<br>THEN email<br>    ELSE 'REDACTED'<br>  END AS email,<br>  country, product, total<br>FROM sales_raw</pre> | <pre>CREATE VIEW sales_redacted AS<br>SELECT<br>  user_id,<br>  country,<br>  product,<br>  total<br>FROM sales_raw<br>WHERE<br>  CASE<br>    WHEN is_member('managers') THEN TRUE<br>    ELSE total <= 1000000<br>  END;</pre> | <pre>CREATE VIEW sales_redacted AS<br>SELECT<br>  user_id,<br>  region,<br>  CASE<br>    WHEN is_member('auditors') THEN email<br>    ELSE regexp_extract(email,<br>'^.*@(.*)$', 1)<br>  END<br>FROM sales_raw</pre> |

Figure 8.5 – Row and column-level selective access along with data masking using regex

In the next section, we will look at another interesting aspect where inadequate and incorrect data sampling, along with human prejudices, may introduce some unfavorable characteristics to influence the insights that are generated, thus making it subjective instead of objective.

# Handling bias and variance in data

We encounter several types of errors in insight generation when using an analytic function. They typically fall into three main categories – that is, bias, variance, and irreducible errors:

- **Bias** is defined as the difference between the "predicted" and "expected" values of an analytic function. The ML algorithm is unable to capture the true relationship between the features and the target. An example of this is model underfitting.

- **Variance** is the result of the model making too many assumptions. An example of this is model overfitting, which means that the training is not generalized enough and should have stopped earlier.

- **Irreducible errors** are random and not directly controlled by the model.

Increasing bias reduces variance and vice versa. In other words, they are indirectly proportional. So, the total prediction error is the sum of all these errors. This can be depicted as follows:

*Prediction error = Bias error + Variance error + Irreducible error*

## Bias versus variance

So, why does bias occur? Let's take a look:

- It could be on account of the human factor because the ML model has been written and trained by a human and is merely mimicking human cognitive bias. NIST's findings from a Harvard study (reference: `https://www.nist.gov/news-events/news/2018/05/nist-study-shows-face-recognition-experts-perform-better-ai-partner`) showed that the accuracy of face technology software in all the top packages offered by companies such as Facebook, Microsoft, Amazon, and Google showed consistently high accuracy for light-skinned males and the least accuracy for dark-skinned females. It is not the software at fault here – it is the training data that was fed to it. The consistent bias demonstrated by all the companies shows a strong human bias.

- It could be on account of poor quality training data since a lot of the deep learning neural network architectures learn from the input data in multiple iterations of trying to minimize the error function. For example, an ML study conducted by Carnegie Mellon showed how quickly a chatbot learned from its less educated audience and believed that the holocaust was fictitious! (`https://www.cbsnews.com/news/microsoft-shuts-down-ai-chatbot-after-it-turned-into-racist-nazi/`)

- Maybe the problem lies in low-resolution unstructured data (image, audio, and video) leading to poorer training and performance.

We can emphatically conclude that bias and variance are both unwanted and undesirable error scenarios. Now, let's look at ways to reduce them.

# How do we detect bias and variance?

Outliers in the data can incorrectly influence the learning outcome. There are statistical methods to help identify them that can then be used to filter out the extreme outliers. Bias in models could largely be attributed to bias in data. There are also tools such as **SHAP** that prioritize the importance of features and help identify instances where the predictions may be off and could lead to bigger concerns: `https://databricks.com/blog/2019/06/17/detecting-bias-with-shap.html`.

Here are a few examples of detecting outliers in data using statistical methods:

- **Standard Deviation** (**STD**): Assuming that the data follows the standard bell curve, if a data point value is over a certain number of standard deviations away from the mean (the default threshold is usually 3), it can be classed as an outlier:

```
import pyspark.sql.functions as F

df_std = (df.groupBy("col1")
            .agg(F.stddev_pop("col2").alias("col2_std"), F.avg("duration").alias("col2_avg")))
df_clean = df.join(df_std, "col1", "left")
            .filter(F.abs(F.col("col2")-F.col("col2_avg")) <= (F.col("col2_std")*3))
```

- **Median Absolute Deviation** (**MAD**): This is based on percentiles and the distance from the median (50% percentile) with a typical threshold of 3:

```
df_mad = (df.groupby(col1)
            .agg(F.expr('percentile(col2, array(0.5))')[0]
            .alias('duration_median')).join(df, "genre", "left")
            .withColumn("duration_difference_median",
                F.abs(F.col('duration')-F.col('duration_median')))
            .groupby('genre', 'duration_median')
            .agg(F.expr('percentile(duration_difference_median, array(0.5))')[0]
            .alias('mad')))

df_clean = df.join(df_mad, "col1", "left")
            .filter(F.abs(F.col("col2")-F.col("col2_median")) <= (F.col("mad")*3))
```

- **Clustering**: Perform the following steps to detect outliers with clustering:

  I.  Run K-means clustering on all the data points.

  II. For each point, do the following:

A. Predict the cluster they belong to.

B. Calculate the distance between the point and the centroid of that cluster.

III. Based on a given fraction, flag outliers

In this section, we looked at some techniques we can use to detect outliers that could contribute to bias in the data. In the next section, we will look at ways to address it.

## How do we fix bias and variance?

One way to fix bias is to ensure that the data is truly representative. We should ensure that the training data is diverse and represents all possible groups or outcomes. For an imbalanced dataset, use weighting or penalized models. Another approach is to do extensive hyperparameter tuning so that the search space is explored properly and an optimal combination is chosen for optimal performance, irrespective of the range of data that's been supplied. Yet another approach is to change the model – for example, a tree algorithm may be better at handling bias. Even though a linear model is simple and convenient, it should not be used if the features and target column do not have a linear relationship.

Now, let's look at how to reduce variance. One way is to train with larger datasets – more data increases the data-to-noise ratio, which reduces the variance of the model. Also, with more data, the model can easily come up with a general rule that will also apply to new data. Another approach is to use ensemble learning – that is, train with multiple models and leverage both weak and strong learners to improve model prediction.

Relying too much on decision support systems can lead us to the trap of automation bias and data is at the heart of it all. Incomplete or incorrect datasets lead to skewed learning. As with every dataset, things constantly change and evolve. So, you may have had a very good representative dataset to begin with but, over time, it degraded. You can only detect this shift if you constantly monitor. To summarize, these errors cannot be eliminated completely but can be largely controlled by some simple strategies, as follows:

- Choose the correct learning model:

  - **Supervised**: Controlled entirely by the stakeholders who prepare the dataset

  - **Unsupervised**: Depends on the neural network itself

- Use the right training dataset:

  - Do not reuse datasets – for example, data from an area with an ethnically diverse population cannot be applied to a region that predominantly consists of a single race.

- Perform data processing mindfully:

  - Not just training, but pre-processing, in-processing (weights), and post-processing (interpretation)

- Monitor real-world performance:

  - Use real-world data to test ML wherever possible.

  - Frequent training.

- Address infrastructural issues:

  - Scrutinize the data collection process.

In the next section, we will learn how to impute data.

# Compensating for missing and out-of-range data

There will be cases where some columns may have missing data. The business use case will determine how serious it is and what to do about it. If a field is being used as an input to a model, it needs a data point. Here are some strategies regarding what you can do:

- Drop the affected records. This is OK when you do not need to use the information for downstream workloads.

- Flag the row/column by adding a marker value (for example, -1). This allows you to see missing data later on without violating a schema:

```
from pyspark.sql.functions import *
avg_age = age_df.where(col('age')!=-1).agg(avg('age')).first()[0]
age_df=age_df.withColumn("age_new", when(col('age') == -1, lit(avg_age)).otherwise(col('age')))
```

- Perform basic imputing so that you have a "best guess" regarding what the data could have been, often by using the mean of non-missing data:

  - The following is an example of filling default values for specific columns:

```
corruptImputed_df = weather_df.na.fill({"temperature": 10,"wind": 10})
```

- The following is an example of using the "average strategy" to impute the values of the specified columns:

```
impute_cols = ['temperature', 'wind']
mean_df = weather_df.na.drop().agg(*[avg(c).alias(c) for c in impute_cols])
mean_weather_df = weather_df.na.fill(mean_df.first().asDict())
```

- Advanced imputing determines the "best guess" using more advanced strategies, such as clustering machine learning algorithms or oversampling techniques (reference: https://spark.apache.org/docs/latest/ml-features#imputer):

```
from pyspark.ml.feature import Imputer
imputer = (Imputer(inputCols=weather_df.columns,
              outputCols=["{}_imputed".format(c) for c in weather_df.columns])
              .setStrategy("median"))
imputer.fit(weather_df).transform(weather_df).show()
```

At this point, we've analyzed the data, addressed its shortcomings, and compensated for some scenarios. In the next section, we will learn how to ensure this pristine condition is maintained over time.

# Monitoring data drift

The famous Greek philosopher, Heraclitus, said "*Change is the only constant in life.*"

Drift refers to the process of moving away from the expected norm. In the world of data, drift is applicable in different contexts. This includes drift in data, in the model, in performance, and in business metrics. Most of the model drift is on account of drift in data. We detect drift in a model by monitoring its accuracy using the F1 score, precision, recall, and other metrics. If the values fall below a certain threshold, then this signals that the business logic needs to be re-evaluated. Drift is usually detected in the context of model drift but that is too late in the pipeline. Profiling the data continuously helps detect drift sooner.

Drift can be classified into two categories, as follows:

- **Data drift**:

  - New fields get added, older fields get dropped or changed, or the statistical quality of the data changes because the product was introduced in a new geography or demography. This causes feature drift and since features are inputs to the model, the generated insights exhibit drift as well. For example, the mean of a value was X but is now Y because a lot of values are empty or missing. Factors such as seasonality, unforeseen circumstances such as the COVID pandemic, and others could contribute as well.

  - Drift in features, labels, and predictions can be seen as an increase in mean and variance.

- **Concept drift**:

  - These are typically external factors that cause the label to evolve. The business definition changes over time – what was once defined as X is no longer true. So, using those rules is no longer valid and the code needs to evolve.

  - This can be either suddenly at a point in time, gradually over time, incremental, or recurring.

How do we detect drift? There are a few core standard scores/tests from standard packages such as SciPy and statsmodel that can help compare older dataset properties with newer ones. Spark provides several APIs to help in this regard:

```python
from pyspark.ml.stat import KolmogorovSmirnovTest

dataset = [[-1.0], [0.0], [1.0]]
dataset = spark.createDataFrame(dataset, ['sample'])
ksResult = KolmogorovSmirnovTest.test(dataset, 'sample', 'norm', 0.0, 1.0).first()
print(ksResult.pValue, ksResult.statistic)
```

The following table summarizes how to handle different features:

| | Numeric Features | Categorical Features |
|---|---|---|
| Summary Statistics | <ul><li>Mean</li><li>Median</li><li>Range (min/max)</li><li>% of missing values</li></ul> | <ul><li>Mode</li><li>Number of unique values</li><li>% of missing values</li></ul> |
| Statistical Tests | - **Mean**:<br>K-S test.<br>It compares the cumulative distribution<br>between the training and post post-training datasets.<br><br>- **Variance**:<br>Levene Test | One way chi-squared test.<br><br>If the p-value is less than the expected alpha, then the expected and observed distributions are different and drift is encountered. |
| Correlation between Feature and Target variable | Numeric Target:<br>Pearson coefficient | Categorical Target:<br>Contingency tables |
| Model Performance | **Regression** Model: Mean Square Error (MSE), Error Plot distributions<br>**Classification** Model: F1 score, Confusion Matrix, ROC | |

Figure 8.6 – Drift detection

`https://github.com/chengyin38/dais_2021_drifting_away`) shows how to use statistical methods to detect drift, while `https://github.com/joelcthomas/modeldrift` shows a model ops framework where it has drift detection as part of its operational pipeline. This ensures that a prescribed automated action can be attempted to rectify the situation:

1.  The first step is to log and version the various datasets and predictions. Delta, with its versioning and time travel capabilities, is handy. Instead of maintaining multiple copies of the data, it is stored in the same table and each batch is versioned and timestamped automatically. We can go back in time to look at the characteristics of past runs and compare them to the current one.

    MLflow is another open source framework that helps manage the life cycle of a model. The creation of each model is logged in detail using the tracking server component of MLflow, allowing for the parameters, code, configuration, and datasets that were used to be fully recorded. This is especially handy in the event of a rollback. We will cover this in more depth later in this book when we address the ML aspects of a data pipeline.

2.  The next step is to apply the statistical tests to the datasets and save the scores.

3.  The last step is to use a visualization library such as seaborn to inspect the data and spot obvious outliers.

Data never lies! So, what story does the data tell and how do we, as humans, interpret that data? Why does data need to be examined and analyzed from multiple dimensions and perceptions? So that we do not tumble down the wrong path of action? This interesting story from WWII will make the point clearer.

Fighter planes returning from combat were evaluated based on the location of bullet holes in their structures. The consensus was to reinforce these damaged areas with more armor – that is, until one statistician astutely reversed the thinking and said that the planes that did not make it back to the base were the ones that were shot at some other spots, so the areas of fortification should be completely different! The planes that got shot could not come back to demonstrate where they were hit, so the existing data from survivors had to be used to imagine where those vulnerable spots were! This is a classic case of survivorship bias and biased views of data interpretation.

# Summary

A modern data format or platform should not only be able to provide for the simple and obvious data flow paths, but also provide strategies for the not-so-straightforward but real-world scenarios that need to be tackled and tamed in production. In this chapter, we explored many scenarios around insufficient and inadequate data and looked at strategies to detect and overcome them. We reinforced the fact that the quality of insights can be controlled by fixing the quality of data instead of over-emphasizing the algorithms that are used to produce the insights. This is because, after a certain stage, it is a case of diminishing returns. However, understanding inherent data issues and fixing them produces a bigger return on the analytic investment. Delta's ACID transaction capabilities, together with its ability to make fine-grained updates, deletes, and merges, allow us to make fixes to the data easily. Everything changes, which means that data patterns, schemas, and drift are expected from all fronts. This is where Delta's time travel and versioning capabilities come to the aid. A good monitoring system is vital to understanding changes to data and proactively responding to data quality degradation to avert bigger ramifications to the business's credibility.

In the next chapter, we will introduce ML aspects to the pipeline and compare and contrast it with just a data pipeline.

# 9
# Delta for Reproducible Machine Learning Pipelines

*"Repetition is the mother of learning, the father of action, which makes it the architect of accomplishment."*

*– Zig Ziglar, American author and motivational speaker*

In previous chapters, we established the pivotal nature of Delta in architecting data pipelines. What about **Machine Learning** (**ML**) pipelines? They involve different personas with different skills and needs. ML has been around for a while; what has changed lately is broad access to large datasets and affordable compute, which has now made it possible for everyone to tinker with ML. Can Delta stand the litmus test of building a reproducible ML pipeline just as effectively as a data pipeline? There are specific challenges and nuances in building a model, staging it in production, and repeating the process over and over again. In this chapter, we will look into these challenges and map the capabilities of Delta to show how it benefits the ML persona through all the steps of feature engineering, model creation, selection, deployment, and monitoring. We will specifically cover the following topics:

- Challenges in ML development

- Formalizing the ML development process

- The role of Delta in an ML pipeline

- From business problem to insight generation

Before going into the challenges of ML, let us look at the role of an ML persona and how we can simplify their day-to-day operations.

# Technical requirements

To follow along with this chapter, make sure you have the code and instructions given at this GitHub location:

```
https://github.com/PacktPublishing/Simplifying-Data-
Engineering-and-Analytics-with-Delta/tree/main/Chapter09
```

Let's get started!

# Data science versus machine learning

Data science is the science of using *data* to solve a measurable real-world problem using a series of *experiments* to perform an in-depth analysis of cause and effect. As in any science experiment, one starts with a question, formulates a hypothesis, conducts several experiments to establish actual evidence, interprets the results dispassionately, and delivers the results, also known as insights, to the business stakeholders who started the chain by asking a question, as shown in the following diagram.

Figure 9.1 – The "science" of data science

There are a lot of steps! This is because data science requires interdisciplinary skills from applied statistics as well as domain knowledge from areas other than software engineering. The core ingredient is data, and there is a whole range of skills that data scientists need to manage and transform it.

Python and R are popular choices for data science, but we are certainly not restricted to just these. In terms of skill sets, there is a wide range, from citizen data scientists who prefer AutoML toolkits, to more sophisticated model developers, and then finally to people who not only leverage existing algorithms but actually create new algorithms. The general trend in the industry is toward the democratization and commoditization of ML, that is, enabling greater adoption of AutoML and API-based model creation, since the science and art behind it has been well established and model interpretability offers a view into the kaleidoscope to examine its efficacy and fairness.

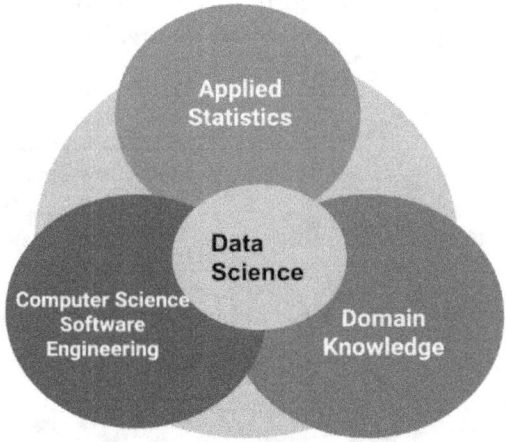

Figure 9.2 – What is data science?

So, if that is data science, then what is ML? Let us find out whether the terms are synonyms or one is a subset of the other. **Artificial Intelligence (AI)** is the process of using a machine to do a task on behalf of a human. ML is a subset of AI where a machine learns from data to make predictions, and **Deep Learning (DL)** is a subset of ML that uses neural networks.

The most rudimentary form of insight generation is to use if/then/else statements and translate a set of business rules into deterministic outcomes. This works up to a point but is not scalable as the data grows and the number of permutations and combinations of rules becomes difficult to control effectively. ML is where a machine learns from data without being explicitly programmed to do so – that is, without being programmed in a rule-based way. An ML system is more probabilistic than that. It may leverage a set of facts, but there is an inherent learning process. It requires more data points and scales well. The following diagram captures the differences between rule-based systems and AI-/ML-based systems.

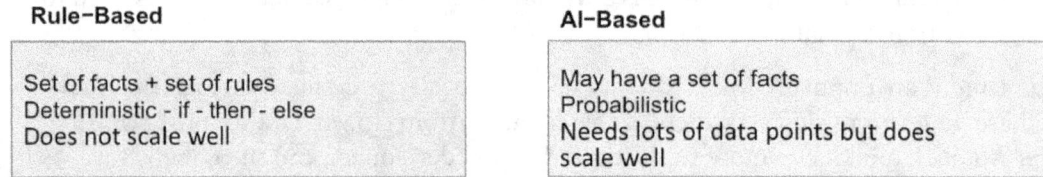

Figure 9.3 – Rule-based versus AI-based systems

In ML, algorithms learn the relationships in data and perform tasks without being explicitly programmed via rules. The raw data in columns is converted into features that carry a lot of predictive power. More data means more wisdom and hence better insights. One thing to keep in mind is that the algorithms need to be computationally efficient to deal with larger data. The following diagram shows the steps in an ML pipeline.

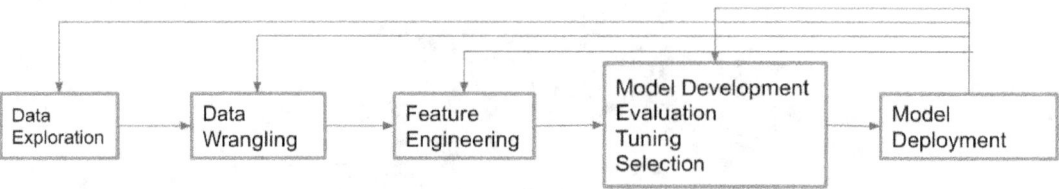

Figure 9.4 – ML pipeline

Now that we have seen the steps involved in data science and in ML, we can better appreciate that ML is a part of data science – it primarily sits in the experiment and analysis phase, as shown in the following diagram, and may sometimes extend into the interpretability and results delivery stage.

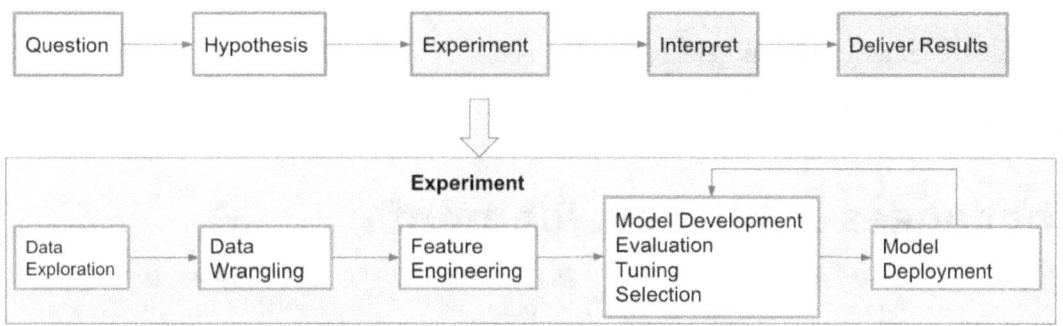

Figure 9.5 – ML is a subset of data science

ML frameworks and libraries fall into the two main categories of classical statistical models and DL/neural networks:

- Classical statistical algorithms:

    - Suitable for most data science problems around structured/semi-structured data

    - Mature and work with low data volumes

- DL/neural networks:

    - Mostly unstructured data.

    - Typically, performance increases significantly with training data volume.

ML personas can be of two types: citizen data scientists and seasoned practitioners. Let us look at each in turn:

- Citizen data scientists:

    - Prefer AutoML toolkits.

    - Look for a fast and easy ML process without many customizations.

- Seasoned ML practitioners:

  - Either leverage algorithms or build their own algorithms to create relatively more complex models.

  - Typically prefer notebook-based environments for exploring and creating artifacts.

Now that we know about data science and ML personas and what tasks they are required to perform in a typical day, let us see what challenges they face at every stage of the pipeline.

## Challenges of ML development

No matter the type of ML persona or the type of model artifacts they produce, the challenges are the same across the board. The following diagram captures the three main challenges around data, people, and tools:

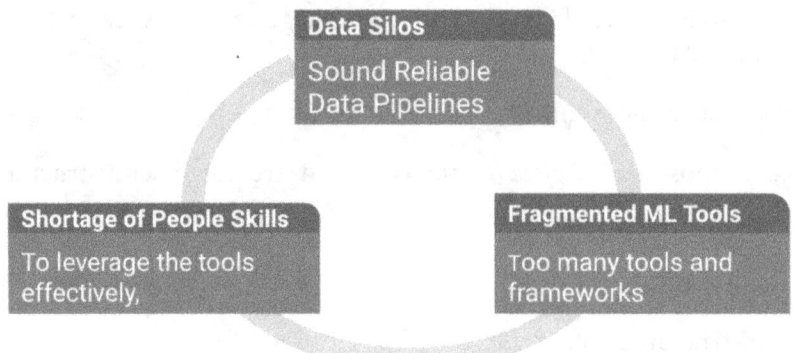

Figure 9.6 – Challenges in ML life cycle management as per Gartner

There is a plethora of ML options to experiment with, but there are only a few people with the skills needed to command the landscape; there are still fewer people who are able to use these tools properly, and even then, their efforts are at the mercy of the quality and completeness of the datasets available to them. Let us examine these challenges in more detail:

- The ML ecosystem is a rich one, with new tools, libraries, and frameworks mushrooming every day. There is a need to be able to experiment with them to check their efficacy without it being too disruptive and distracting. In other words, ML practitioners need the ability to layer in a new flavor without affecting, say, the data cleansing and transformation aspects upstream, or the DevOps activities downstream.

- A lot of experiments have to be conducted to find a model suitable for production. What are *all the variables* that need to be tracked and recorded to capture the experiment with full fidelity so that it can be reproduced later either in the same or a similar environment? There are times that an ML persona has stumbled on a great model that has been running on a laptop and cannot make its way to production because all the details around how it was generated are not known or fully documented. How often have we heard the blame-shift story of "Worked fine in dev, and since it doesn't work in production, it must be an ops problem." Such instances result in a single point of failure and are wasteful. Here are some of the variables to be considered:

  - The **data** used to train the model.

  - The underlying **code** and choice of model architecture.

  - The **dependent** libraries and their versions, so that you can recreate the environment.

  - The **features and hyperparameters** involved in that specific run.

  - The **model artifact** and any associated assets, such as weights and AUC/ROC charts. There could also be descriptions, notes, tags, and other metadata, such as who created the model and how long it took to train.

  - The relevant model **metrics**, such as accuracy, precision, recall, and F1 score, that could be used to compare against other models created from the same dataset.

- A model that has made it into production is a wonderful achievement! However, the journey does not stop there. There is a whole separate pipeline around **model management**. Over time, the model becomes stale and needs to be retrained. Yet another separate pipeline to **monitor drift** is needed. Model drift is often on account of data drift and is a signal to trigger a retraining process. This is where the champion model in production is compared against a new challenger version to see whether it is time to be replaced or not. Over time, it is important to be able to query what version exists in production, so that there is no confusion about which is the active one, which is the challenger, and which one needs to be promoted or rolled back. Many people have no idea what version is in production! This is where a central **model registry** that serves as the single source of truth for the models and their stages and versions is imperative.

- In regulated industries, it is important to show a **transparent audit trail** of what model was used to score certain critical business functions for end users to ensure fairness and compliance. These data trails are all orthogonal to the model creation step.

- Feature engineering is a compute-intensive task and needs to be done carefully as the quality of the data features fed to models determines how good the model is. These features are created once and reused across multiple use cases. Moreover, the same features are used at inference time, and this helps eliminate the online/offline skew problem. So, having a good **feature store** that maintains multiple versions of the generated features and keeps it relevant with not only the features but details on the lineage of how the features were derived is important for both the producers and the consumers of these features.

- **Model Interpretability** and governance to ensure fairness and soundness of the model is another important consideration. Any model can produce some output but the ability to justify the outcome is especially important in regulated industries. It sounds clichéd, but it's apt to say, "*Data, data, everywhere, and not a drop of insight!*" The governance challenges include the following:

  - Can you justify the best model-selection process for production?

  - Can you reproduce the model with absolute fidelity?

  - Can you detect model drift and replace a model with a newer version?

  - Do you even know which version is currently being served in production?

ML is a very iterative process and involves multiple data personas. Often, the hard part of ML is not the ML itself but rather the *data* and *process* parts. If ML is hard, then productionizing ML is harder. **Continuous Integration/Continuous Deployment (CI/CD)** has to be part of your ML story if you are serious about ML.

The ML life cycle is more complex than just a vanilla software development life cycle, so let us see how to formalize the process to reduce risk.

# Formalizing the ML development process

Let us first understand what constitutes a model and how we define MLOps. It is important to emphasize that this chapter is not about the nitty gritty details of creating a model, but rather about the data aspects of creating a good model and the process of continuously refining it to keep it relevant and useful to the business.

# What is a model?

A model is an artifact that has several inputs and outputs. Let's list them so we have a firm idea of what a model encompasses. The following diagram captures our definition of an ML asset and its refinement zones:

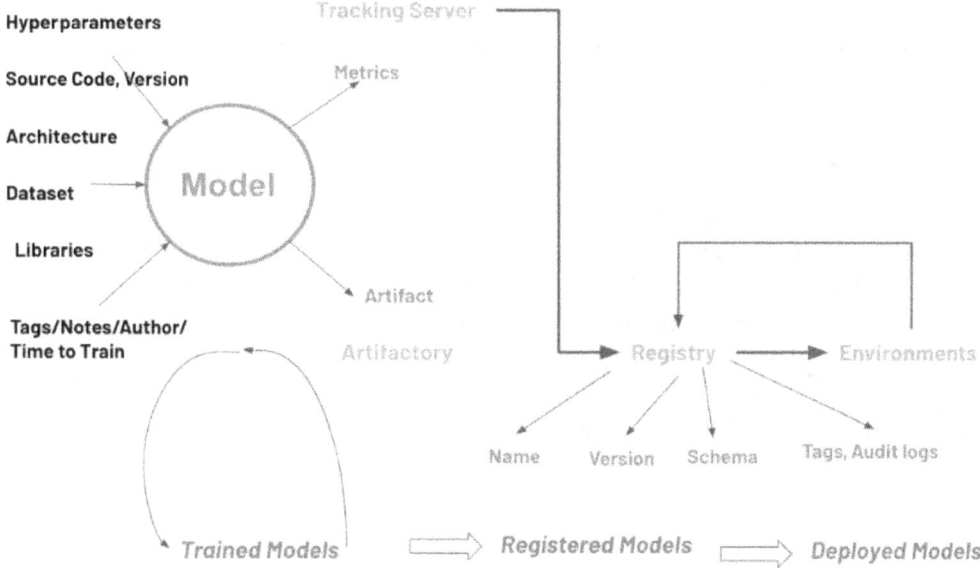

Figure 9.7 – What is a model?

The inputs to this process include the following elements:

- One or more datasets
- One or more libraries used to create the model
- The source code used to create the model with a given architecture
- The distinct values for the various hyperparameters used to train the model
- Additional metadata such as author, tags, time to train, business use case, justifications, and any other notes

The outputs of this process include the following outcomes:

- The model artifact, such as a pickled file along with artifacts including weights, graphs, and logs
- A set of model metrics that a stakeholder will use to examine the model efficacy

When talking about data pipelines, we looked at the medallion architecture of bronze, silver, and gold for curating and refining data. In the model world, these refinements are captured incrementally in a *trained* model, a *registered* model, and a *deployed* model.

There may be thousands of trained models, of which one is selected as the best to be used as a registered model. A model in a registry has some attributes, such as a name, version, schema, tags, and other descriptive details. It is also a good place to track the version of the models in the various environments. There may be several registered models but typically just a single deployment model.

## What is MLOps?

MLOps is referred to by various names but is the intersection of data engineering, ML, and DevOps and refers to an automation process where you design a pipeline, train it and run it. It is a shift from a **model-centric** view to a **data-centric** view of insight management. We can think of it as an additive equation:

$$MLOps = ML + DEV + OPS$$

Let's visualize the intersection of data engineering, ML, and DevOps:

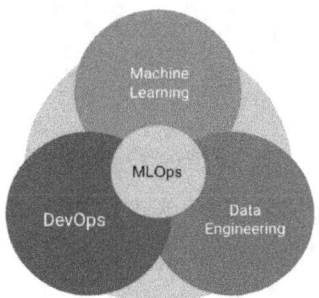

Figure 9.8 – MLOps

Every organization is at a different level of maturity in its MLOps journey, but the important thing is to get started and continuously improve and expand to reach newer heights of maturity.

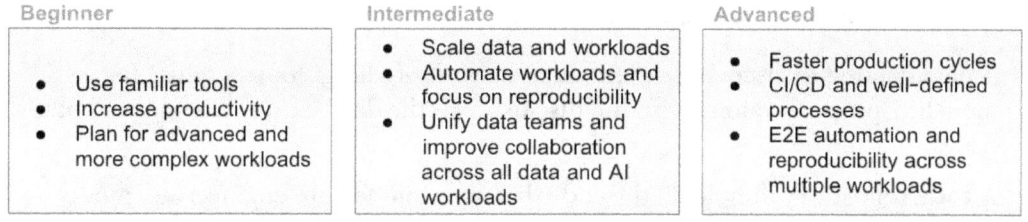

Figure 9.9 – MLOps maturity

MLOps involves making progress in multiple key capability focus areas, including people, data architecture, model architecture, process, and governance. The various components of a robust MLOps strategy include **Exploratory Data Analysis (EDA)**, data preparation and feature engineering, model training and tuning, model review and governance, model inference and serving, model monitoring, and automated model retraining.

# Aspirations of a modern ML platform

Based on the challenges listed in the previous section, the following are the aspirations of a modern ML platform.

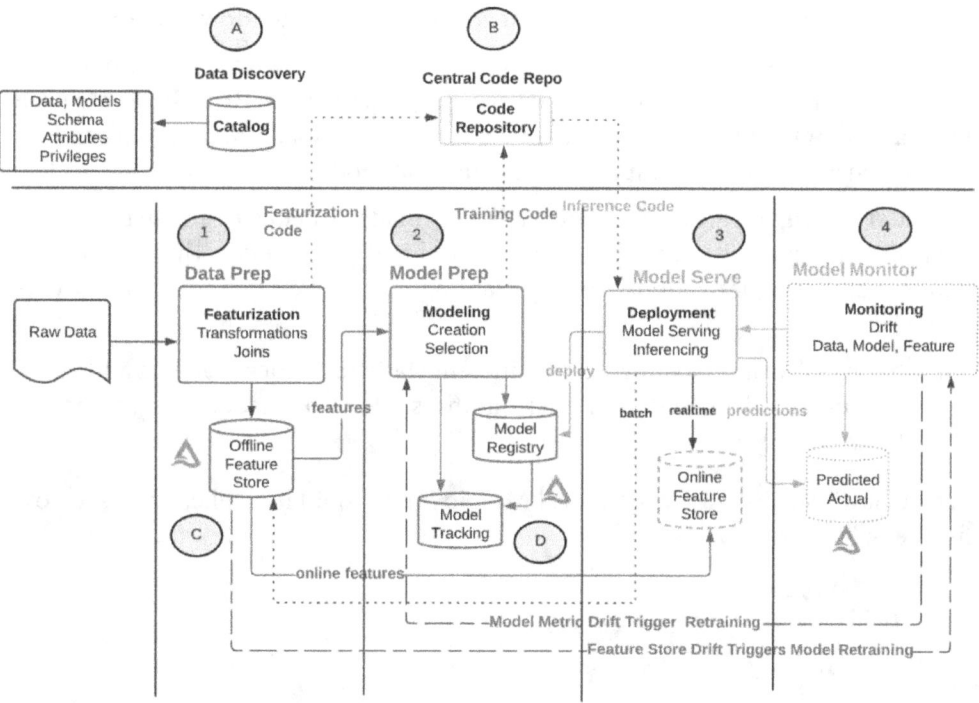

Figure 9.10 – Formalizing the ML process

The four main dependent components include the following:

- A **data catalog** to discover relevant datasets with predictive power to use for modeling purposes, along with full disclosure of the data source, schema, and other relevant metadata.

- A **code repository** to hold all the code that goes into feature engineering, model creation, inference generation, drift detection, and pipeline orchestration.

- A **feature store** to hold the engineered features. It may have an optional online feature store component for use in real-time predictions.

- A **model management** component to ensure the path to production is repeatable. Model tracking holds all the results from the various runs of an experiment while creating a model, and the model registry tracks the chosen model along with its various versions (as it moves across various environments such as development and staging before making its way into production), where it is discoverable to others for reuse, much like the data catalog.

The four main pipelines, which work in a symbiotic fashion, include the following:

1. A **data preparation** pipeline where the features are engineered and the computed values are formally made available in the feature store for use downstream.

2. A **model preparation** pipeline where a plethora of ML tools and frameworks, including AutoML frameworks in multiple languages using CPUs/GPUs, are used to create/train the model artifact. The best artifact is pushed into a registry to be discovered and used in multiple use cases for that model.

3. A **model serving** pipeline where the artifact is made available in a target environment, where it is used to score data and make inferences. The various deployment strategies include batch, streaming, RESTful endpoint, and deployment at the edge.

4. A **model monitoring** pipeline where different statistical aspects of data and model are monitored for drift, and once a certain threshold is violated, it triggers the model retraining process.

Now that we have established what an ideal ML pipeline should look like, let us see how Delta fits the bill.

# The role of Delta in an ML pipeline

Delta's capabilities around ACID transaction support, schema evolution, and time travel come in handy in the context of designing ML pipelines. Let us examine details of each of the four co-operating pipelines involved in creating and managing an ML asset.

## Delta-backed feature store

Feature engineering is time-consuming and involves resource-intensive computation, domain knowledge. Poor feature engineering can have an adverse impact on the quality of ML models, so a lot of attention and care should be given to its computation.

Features are the inputs to ML models and they have to be computed based on raw data. Feature augmentation and pre-computed features require a feature store that precomputes those features and makes them available both at training and serving.

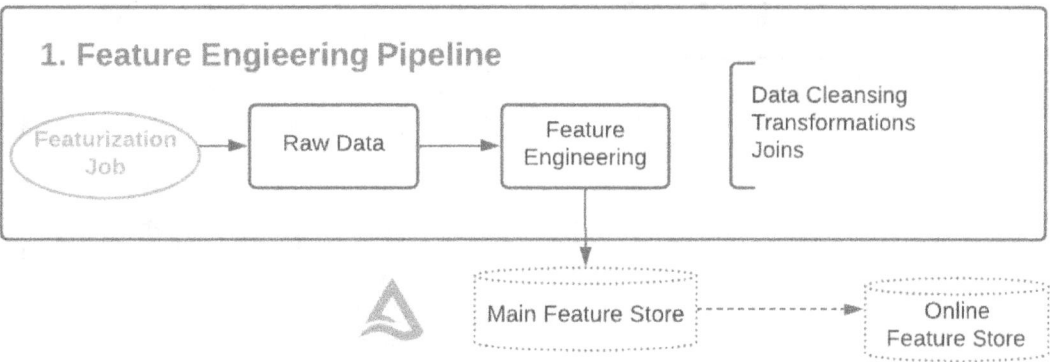

Figure 9.11 – Feature engineering pipeline

Features can be of several types, such as transformative which requires category encoding, context features such as weekday versus weekend, additional augmentation such as weather, and geospatial details. There might be pre-computed features based on a business logic such as a customer's purchases made in the last week, month, or year, in order to predict the probability of a customer making their next purchase.

The requirements for a feature registry would include the lineage of transformations, the versioning of generated features, and discoverability and reuse across multiple use cases as well as between training and inference instances to avoid the problem of online/offline data skew. Delta, with its open format, built-in versioning and governance, and native access through a variety of APIs backed by Python and SQL, makes Delta an ideal choice for a transactional feature store. It can have multiple provider interfaces; for example, for batch storing large volumes of data with SLAs of several minutes, Delta can be used from the main store. It also has the ability to share data securely with a different online store for real-time scoring with more stringent SLAs. These features have built-in time dimensions by virtue of Delta's time travel capability, which allows the right set of features to be joined appropriately to incoming data based on its time properties.

Managed offerings of the feature store simplify this by providing APIs to interact with the feature store.

| Before | After |
|---|---|
| ```
raw_data = spark.read...
features =
raw_data.select(...).agg(...)...
...
features.write(...)
``` | ```
fs = FeatureStoreClient()

def compute_features(raw_data):
    features =
raw_data.select(...).agg(...)...
    ...
    return features

fs.write_table(...,
compute_features(...))
``` |

Figure 9.12 – Example of using the Databricks Feature Store API

Let us look at the next pipeline where the model is trained.

# Delta-backed model training

An ML practitioner experiments with several model architectures and hyperparameter permutations to hone in on a selected model. Different practitioners may collaborate on this effort following a divide-and-conquer approach. In order to compare apples to apples, the models need to be created using the same dataset irrespective of whether it was created by one or many individuals over a certain time period. The data in the underlying datasets is constantly getting refreshed, added to, and otherwise changed by various people. This would lead to inconclusive comparisons. Usually, people take a copy of the data so that it remains pristine, but if a feature is changed, that requires yet another copy of the data, and very soon this gets out of control. Delta's time travel capability eliminates the need to maintain multiple copies of the data. Practitioners can agree either on a timestamp or version of the data and all of them will get the same view of the data and can proceed independently. Instead of `SELECT * from <table>`, they can use `SELECT * from <delta table> AS OF Version N`. The following diagram shows the addition of a second pipeline that is responsible for the creation and selection of the model, and it takes the features generated from the earlier pipeline as one of its inputs. This capability is also handy when reproducing a model. As noted earlier, one of the main inputs is the dataset used to train, which is a constantly moving target. However, noting the version of the raw dataset or feature store table version or timestamp with the model statistics will ensure that we can get the same data points to recreate the model at a future point in time, maybe in a different environment such as staging or production.

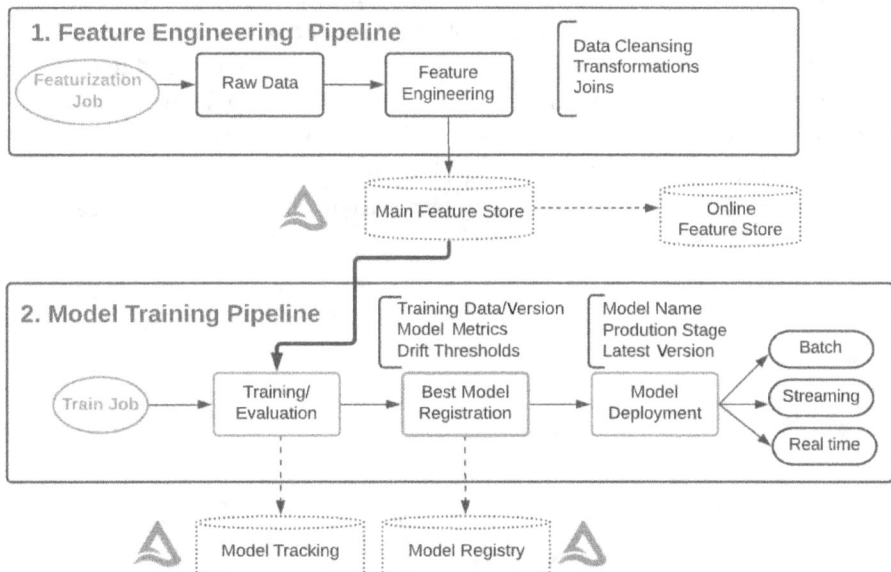

Figure 9.13 – Adding the model training pipeline

Each run of the experiment can be logged in a Delta table and the final model can be put in an artifactory with a link to its location in a Delta table. On a managed platform such as Databricks, these come out of the box as the tracking server and model registry MLflow components, which encapsulate all the necessary details and expose the functionality through APIs. Delta Clone can be used to make a deep copy of the datasets used for training and validation to safeguard against periodic vacuum operations performed on the table over time. Model selection is a good point at which to further formalize the experimentation process and optionally create a copy of the data for permanent archival. The model deployment is orthogonal to the creation process and different flavors of the model can be deployed in different environments to meet varied consumption patterns. This supports batch, streaming, and real-time inferencing, the details of which we will cover in the next section.

## Delta-backed model inferencing

As new data arrives to be scored, the third pipeline pulls the relevant features from the feature store, uses the right version of the model from the model registry, scores the data with it, and saves the results of the prediction into a Delta table. In some use cases, the ground truth is known immediately; in others, it may take some time and trickle into the system later. The ground truth refers to the "actual" behavior. For example, say an ML inference recommended a product to a user and the user clicked on it. This click information demonstrates that the recommendation was valid and of interest to the user. We can conclusively state that the predicted value and actual outcome were verified. However, if the recommendation was instead offered as an attached coupon to a pharmacy receipt, it might take several weeks before the user returns to the store and uses it, if they ever show up again. This scenario of collecting the actuals is a longer process and the accuracy of the prediction cannot be determined till the ground truth is made available. When it does come, Delta's ability to do fine-grained update and merge operations for the existing data helps bring the two data points together to determine model efficacy.

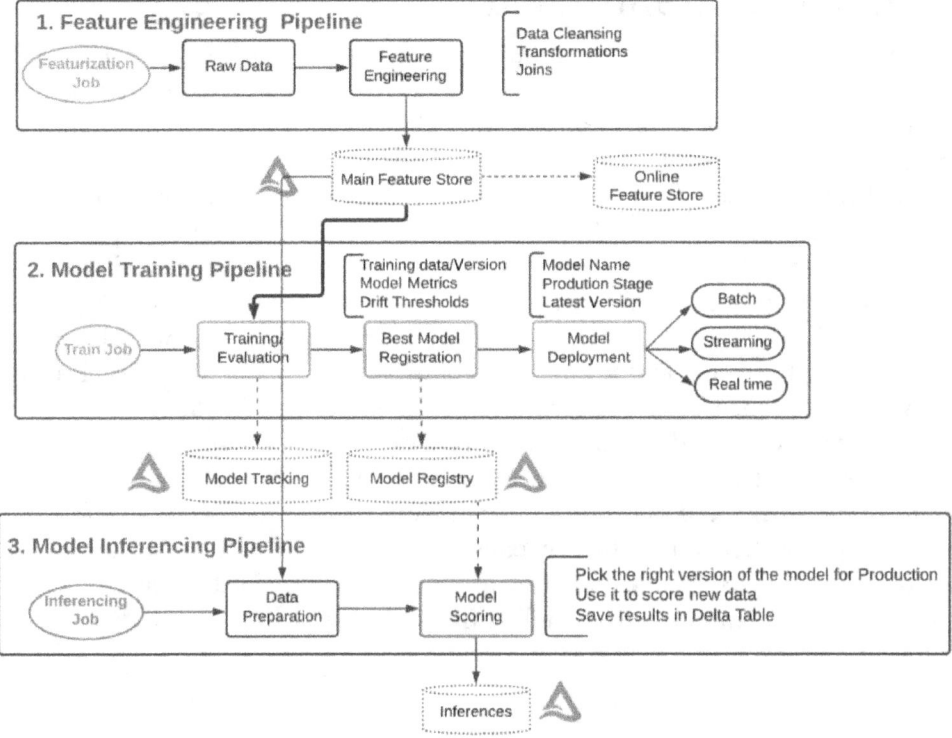

Figure 9.14 – Adding the model inferencing pipeline

ML creation and management is an iterative process and does not stop at inferencing.

# Model monitoring with Delta

Many argue that model drift is best monitored by monitoring the data drift in incoming data and the drift in the generated features. As and when the ground truth is available, it is joined by some primary key criteria with the inference data in a Delta table. Again, the update and merge operation support in Delta makes this a breeze. Now the actual and predicted values of the inference data are computed to see how well the model is doing in terms of the quality of insight generation. The feature engineering pipeline is completely in-house and is easier to monitor for drift. The model interpretability may indicate that some columns contributing to the predictive power are incorrect, and it may be necessary to add or remove features. In such cases, a threshold of tolerance is violated, which signals a need for model retraining. Each of these four pipelines runs on its own job schedule, with some dependencies in their inputs and outputs, as shown in the following diagram. So, the detection of drift can generate an alert to let a DevOps or ML persona know that there is a potential change to be investigated, or it may directly invoke a job for the training of the pipeline, so that a fresher champion model is made available for A/B testing. It is also important to note that schema changes to input data could lead to new features being engineered and added as inputs to the model. Delta's schema evolution features come in handy for handling this change gracefully. The four pipelines are as follows:

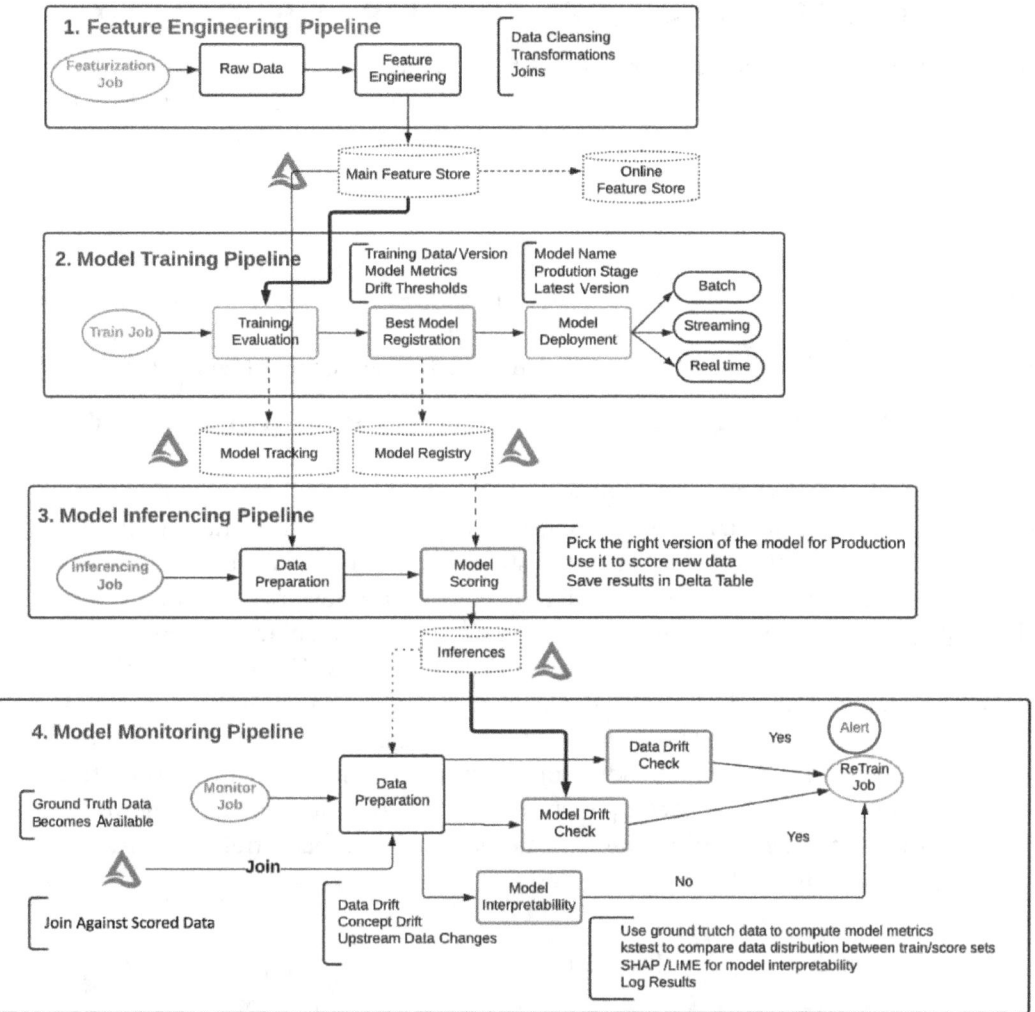

Figure 9.15 – Adding the model monitoring pipeline

A managed platform such as Databricks simplifies a lot of this by providing out-of-the-box features to leverage so that you do not have to build them from scratch. For example, the model serving and model monitoring are infrastructure components that would add value to an ML initiative, allowing the practitioner to focus on the relevant pieces around the use case, and the platform can abstract away all the other complex integrations.

# From business problem to insight generation

No data science or ML is justified purely by tech; there needs to be a business problem at hand. The business team usually comes up with a prioritized list of problems to solve. The consumers could be an internal or external audience. Dashboards are a great way to communicate insights. The key metrics are constantly tracked and transparently available for everyone in the value chain.

Let us look at the four cooperating pipelines discussed in this chapter through the lens of ownership and handoffs. The skills, tools, and frameworks that these personas wield are sometimes non-overlapping, so it is interesting to consider how they collaborate with one another to create the final product or service and satisfy their stakeholders in the business.

It is fair to say that the data engineer is responsible for the **Extract Transform Load** (**ETL**) process and bringing in different data sources. However, they may not have much domain knowledge, so it is usually the data scientist who does EDA to understand the statistical importance and quality of the data, and they may work with the data engineer to do the feature engineering and generate interesting features for the feature store for reuse downstream. A citizen data scientist may use an AutoML framework to create a model while a sophisticated ML practitioner may create one from scratch, try several variants, and select the best one for further testing in later environments. The MLOps engineer has software engineering and ML skills and steps in to perform a lot of the automation and integration with CI/CD systems, final testing, and packaging to get ready for the green light to push the model into production. Inferencing on new data with the latest model from production is usually completely automated. The monitoring framework reports on how the model is performing and alerts on drift so that retraining activities can commence, as shown in the following diagram:

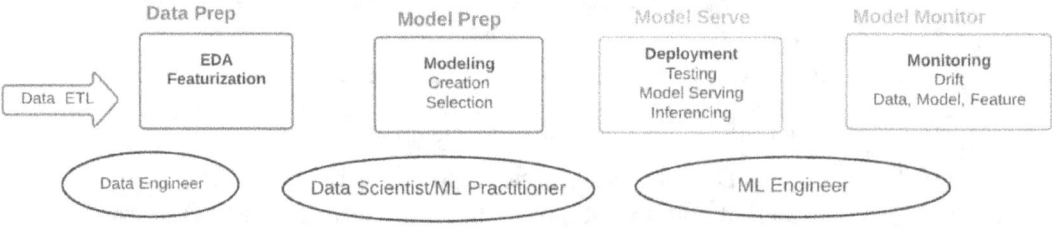

Figure 9.16 – Handoffs between multiple personas during ML asset creation and management

The definitions of the roles and responsibilities should be taken with a pinch of salt, as people wear several hats; a single person might be running the whole show, or there could be more division of labor among the main personas. One aspect of governance that should be noted is that there may be some elements of sensitivity around the data, features, and model, and care should be taken to manage the access privileges so that nothing falls into unintended hands. The same is true of the promotion process. The approval process may tie people, processes, and tech together. The automated test results are the contribution of the tech; the workflow of the required steps and dos and don'ts are part of the process that may be unique to an organization or use case; and the people are the folks with the authority to approve or deny the process based on their judgment of the data presented.

# Summary

Insights generated from ML models provide a competitive advantage to a business. However, the process is complex and there is a certain level of discipline that needs to be followed to maximize the return on investment. There are certain core components, such as a feature store, a model registry, a code repo, and a catalog, that are necessary to streamline the ML process, as it is very repetitive and it would be a shame to waste the valuable time of data scientists for tasks that are removed from the use case at hand. The model management aspects cannot be ignored either, because once created, an ML asset is a living, breathing entity that needs care and attention to ensure that it is performing as expected.

In this chapter, we looked at Delta through the lens of an ML practitioner and examined how it adds value to their day-to-day operations on several fronts, including feature engineering and reuse, model training with a unified view of the dataset, model reproducibility referencing the same dataset, model inferencing leveraging the feature store and saving inference results into a Delta table, and finally detecting model drift and maintaining the model monitoring results in a versioned Delta table to analyze and compare over time. In the next chapter, we will look at how Delta facilitates the creation of data products and offers data as a service.

# 10
# Delta for Data Products and Services

*"At the heart of every product person, there's a desire to make someone's life easier or simpler. If we listen to the customer and give them what they need, they'll reciprocate with love and loyalty to your brand."*

*– Francis Brown, Product Development Manager at Alaska Airlines*

In the previous chapters, we saw how Delta helps not only with data engineering tasks but also with **machine learning (ML)** tasks because data is the core of all ML initiatives. **Data as a Service (DaaS)** refers to data products being made available to users on demand. The popularity of microservices and APIs has facilitated access to data on a need and privilege basis. These users can be within the organization or outside vendors and partners. The advantage is that the consumption pattern is greatly simplified and standardized and the internal complexities of pipelines and data stores are abstracted away from the end user. The data can be at different levels of refinement. For example, it could be the final insights generated from statistical computations and ML prediction insights. This offers additional monetization opportunities to the organization that can anonymize data for general consumption.

A lot of the data around us remains in an unstructured form and carries a wealth of information. Integrating it with the more structured transactional data is where firms can not only get competitive intelligence but also begin to get a holistic view of their customers to employ predictive analytics. For data to be truly democratized, it needs to be prepped for easy sharing as a product and a service. Prepping may involve data mashups collating different types of data from different sources. In this chapter, we will look at how Delta offers first-class support for unstructured data and offers secure, scalable sharing options.

The following topics will be covered in this chapter:

- Data as a Service (DaaS)

- The need for data democratization

- Delta for unstructured data

- Data mashups using Delta

- Facilitating data sharing with Delta

Let's look at what the typical use cases are and how organizations set up data exchanges to facilitate the creation and consumption of data products and services.

# Technical requirements

To follow along with this chapter, make sure you have the code and instructions as detailed on GitHub here:

```
https://github.com/PacktPublishing/Simplifying-Data-
Engineering-and-Analytics-with-Delta/tree/main/Chapter10
```

Let's get started!

# Data as a Service (DaaS)

Every company wants to be data-driven but just collecting data doesn't make you a data-driven enterprise. However, having actionable customer-centric insights does. If there is a data provider that can do the heavy lifting for easy consumption, then every business can use its services to be more competitive and truly data-driven. Being a DaaS provider is hard because taming raw data is a herculean task and it takes a lot of planning and execution to pull it off. The typical activities to manage data include the following:

- Data collection to ensure quality and timely data

- Data aggregation to summarize data in well-known dimensions for actionable insights and to avoid analysis paralysis

- Data correlation for proper data and risk modeling to use the predictive value inherent in the datasets

- Qualitative analysis to ensure that there is statistical significance and insights generated from the data that can be relied upon

- Advanced BI and AI analytics to provide easily digestible nuggets of information that can be put to use immediately

You may say, "Well, this is only relevant to large organizations." That is not true. Enterprises of all sizes can benefit from this. Take the following examples:

- A new business venture needs data points to prove gut reactions to the market and make a proposal to an investor. DaaS helps by providing deeper insights to innovate faster and minimize risk.

- A mid-size company needs that analysis to remain competitive.

- A larger organization continues to analyze alternate data, particularly unstructured, to break down data silos and ensure that the current operational strategies are optimum.

In some ways, DaaS is a value framework to monetize big data. Some of the key benefits are listed here:

- Improved product and service offerings, leading to optimal operational workflows.

- A better understanding of the customer, leading to improved customer experience, increased retention, and loyalty.

- Higher ROI and better pricing strategy, leading to better customer lifetime value.

There are several roles to be played by data personas; the main ones include app development, advanced analytics, domain experts, data management and admins, system integrators, and software engineers working alongside creative personas to collect, wrangle, and present the data. In the next section, we will explore different use cases across industry verticals to emphasize the need for data democratization.

# The need for data democratization

Data democratization refers to the process of making data available to all relevant stakeholders to consume as is or add further value. This is critical for all businesses as it forces agile, data-driven decision making and helps them to remain competitive using actual metrics and data-centric strategies, as well as providing monetization and innovation opportunities. Let's take a look at a few concrete examples:

- **Healthcare and manufacturing**: A new category of medical imaging device has been introduced into the market. A lot of vendors and hospitals buy these devices. The images, their quality, and their predictive power in aiding doctors to detect the onset of tumors and cancers based on certain positions and circumstances generate data points that need to be analyzed to see what positions and settings lead to the best diagnosis. The more data, the better the analysis and the quicker the feedback loop to provide to the manufacturer to improve the product or provide better-recommended settings to get set up correctly right from the get-go. This applies to any product feedback collection over a larger surface area.

- **Fraud and compliance agencies**: A financial institution is expected to report on **suspicious activity reporting (SAR)**. This will help authorities to see the same pattern elsewhere, correlate the two, and prevent potential fraud activity proactively.

- **Financial services**: Crypto is on the rise and every financial institution is trying to ingest the same data from multiple exchanges, curate it, and analyze it to provide insights to its retail and institutional investors. There is one part of the data that is a fact (ground-truth transaction), which each consumer should use as is and then add their special sauce to to get superior insights and remain competitive. A good example of this is the Google Search engine, which crawls millions of documents to provide curated relevant search results. For a time, there will be a lot of folks playing the role of provider. Eventually, the credibility and quality of the data will lead to the establishment of the top leaders. So, a Bloomberg or Google equivalent could take it upon itself to provide this data to all financial institutions or individual users.

These are all good examples of breaking down data silos and unnecessary repetitive tasks that help lower risks associated with incomplete or poor data quality, while lowering costs on a task that needs to be done once. The following diagram illustrates the constant flow of data from producers to consumers, and any efficiency that can be built to facilitate the process brings about several-fold acceleration of the insight generation workflow:

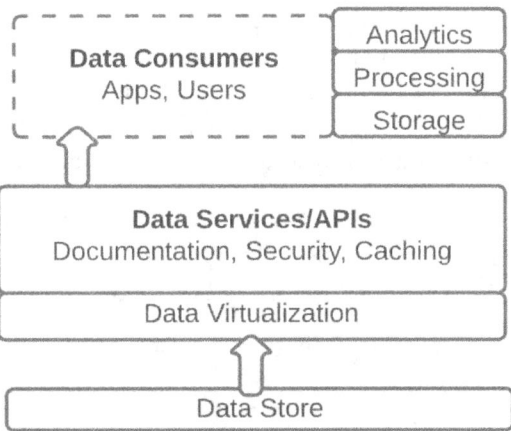

Figure 10.1 – Data producers and consumers

What happens when we are dealing with large datasets or a curated dataset that is meant for certain individuals? Moving around large datasets is expensive, time consuming, and error prone. Data sharing without making a copy of the data is at the core of bridging the producers and consumers that facilitates the data democratization process. Let's look at whether DaaS is a viable option for the secure transfer of large datasets. This is further exacerbated when we are talking about unstructured data, which is more complex to store, process, and analyze.

The other related concept we hear about is that of **data as a product** (**DaaP**) in the context of a data mesh, where functionality is divided by business domains that are entity driven.

*"A product that facilitates an end goal through the use of data"* is how DJ Patil, the former US chief data scientist, referred to a data product in his book *Data Jujitsu: The Art of Turning Data into Product*, 2012.

*"Domain data teams must apply product thinking to the datasets that they provide; considering their data assets as their products and the rest of the organization's data scientists, ML and data engineers as their customers"* is how Zhamak Dehghani from Thoughtworks articulated a data product in the context of a data mesh (this quote is from the podcast transcript available at `https://www.infoq.com/podcasts/ domain-oriented-data/?itm_campaign=popular_content_list&itm_ content=&itm_medium=popular_widget&itm_source=infoq`). The required capabilities include data discoverability, adherence to security rules, the ability to explore and understand using a catalog or meta store, and trustworthiness, which translates to data quality and credibility. Each of the boxes in the following diagram is a domain-specific data product with some autonomy and they share insights with one another:

Figure 10.2 – Data products contributing to the data mesh

The three main characteristics of domain-specific data mesh architectures include the following:

- **Product thinking**

  - Each domain-specific group offers its wares as data products to the rest of the organization in a manner that is discoverable, interoperable, and trustworthy.

- **Distributed architecture**

  - Focus on a pull model where each domain team is responsible for designing, building, and maintaining their own ETL process

- **Designed to facilitate self-service**

  - The common infrastructure aspects across all domains are abstracted away and generalized to provide reusable components to onboard new data products.

It is important to contrast this with a monolithic design, which slows down innovation as centralized platform teams start to become the bottleneck. The other extreme is siloed business units where there is a lot of duplication of work, leading to increased costs and resulting in technical debt, which could easily have been handled by borrowing from more mature parts of the organization. Some organizations are somewhere in the middle, relying on the **center of excellence (COE)** or central platform team to bring in their data as a unified framework, after which the business units take over. This is closer to the mesh framework but does cause some unnecessary dependencies on a COE team. To reiterate, in the mesh framework, the COE is tasked primarily with governance, providing base infrastructure automation scripts and up-to-date documentation and enablement material to allow data teams to be agile and self-sufficient, allowing data to indeed be a first-class citizen. Another key difference is in the team composition. Instead of role-focused personas, data mesh teams are cross-functional and domain oriented. Instead of data artifacts, it is data itself that is the first-class citizen!

Now, we have two terms, DaaS and DaaP, but what is the difference? It is fair to say that your data pipelines in individual domains refine data and create data assets, such as dashboards and models, which are the data products, a combination of them serves a more strategic business use case, and the access to data is facilitated using a data service. From a maturity perspective, data products blossom over time into mature data services. The data products just facilitate access to company data with a focus on SLAs and access privileges. The data service, on the other hand, munges data with possible external data sources and the consumption pattern is optimized by not only the engineers, analysts, and ML practitioners but also the product, marketing, and other business folks. Also, the data serves higher predictive answering patterns in a partnership-style relationship, for example, between a company and its vendors.

This section focused on collaboration and synergy across different businesses using holistic data for their specific use cases while still retaining some levels of isolation and autonomy. In the next section, we will look at unstructured data, which is analogous to the large portion of the iceberg below the surface of the water, as it is not directly visible but contributes essential signals to augment the structured data elements visible above the water.

# Delta for unstructured data

The vast majority of data in the world is unstructured – estimated by analysts to be 80 percent of all data they generate or otherwise acquire while doing business. Video, audio, or image files, as well as log files, sensors, or social media posts, all qualify as unstructured data and it is growing at a faster pace than structured data. Object storage technologies have facilitated the storage of all data types in a cheaper, more scalable, and reliable manner, and this has largely been responsible for the increased support of a large variety of use cases. This has led to a spike in deep learning models. Typical use cases include the following:

- Image classification
- Voice recognition
- Anomaly detection
- Recommendation engine
- Sentiment analysis
- Video analysis

Spark supports the `image` format as well as the `binary` format. The image format has a few limitations around decoding image files during the creation of the DataFrame, leading to an increase in data size, which has undesirable consequences, and so the recommended format is binary. Each binary file is converted into a single record that contains both the metadata as well as the raw content. The fields include the following:

- `path`: The file path
- `modificationTime`: The time stamp
- `length`: In bytes
- `content`: Binary content being ingested

Let's now look at how Spark can be used to ETL unstructured data, starting with the ingestion of binary files, transforming it, and persisting it into Delta tables:

1. **Extract**: Ingest data to read a set of images in a given directory either in batch or streaming mode. To provide additional annotation, you can use the `mimeType` option, as shown here:

```
df = spark.read.format("binaryFile") \
            .option("mimeType", "image/*").load("<path-to-dir>")
```

To recursively search files, we can use additional options, as shown here:

```
df = (
    spark.read.format("binaryFile")
    .option("pathGlobFilter", "*.jpg")
    .option("recursiveFileLookup", "true")
    .load("<path-to-image-dir>")
    )
```

2.  **Transform**: After ingesting the data, there may be a need to extract some metadata, such as the size from the image by using a **user-defined function (UDF)** to load each image and pull its size:

```
def extract_size(content):
  #Extract image size from its raw content
  image = Image.open(io.BytesIO(content))
  return image.size

@pandas_udf("width: int, height: int")
def extract_size_udf(content_series):
  sizes = content_series.apply(extract_size)
  return pd.DataFrame(list(sizes))

df = images.select(col("path"), extract_size_udf(col("content")).alias("size"), col("content"))
```

3.  **Load**: Data can be persisted in batch or streaming mode, as shown in this example:

```
(df.write.format("delta")
 .option("checkpointLocation", ckpt_path)
 .trigger(once=True)
 .start(delta_path))
```

Now that the data is available in Delta format, users further downstream can use any language, including SQL, to analyze the data:

```
df_saved = spark.read.format("delta").load(delta_path)
df_saved.createOrReplaceTempView("tmp_flowers")
```

The binary format can be used to read not only image data but also other binary formats, such as audio, video, or text. For example, to read audio files with the .wav extension, we can use the following:

```
(df.write.format("delta")
 .option("checkpointLocation", ckpt_path)
 .option("pathGlobFilter", "*.wav")
 .trigger(once=True)
 .start(delta_path) )
```

Let's next look at **natural language processing** (**NLP**) and image data types as they are the most popular in the unstructured world.

## NLP data (text and audio)

NLP involves use cases where the intent/meaning is derived from text and audio data. People type inquiry emails and complaint notes and interface with call centers using audio, which generates vast amounts of unstructured data. Audio data is often converted into text and then analyzed. Typical scenarios include the following:

- **Sentiment analysis**: Analyzing the sentiment of customer product feedback
- **Question answering**: Answering questions based on an article
- **Machine translation**: Translating from one language to another
- **Chatbots**: Chatbots for answering customer questions or requests
- **Automatic summarization**: Summarization of news articles or various tweets about a common subject

NLP can be used with single-node ML libraries, such as NLTK, spaCy, TextBlob, Gensim, and Stanford CoreNLP. However, to do it at scale, distributed libraries, such as SparkML, Spark CoreNLP, and John Snow Labs Spark NLP, are available as well. It is the open source storage layer that brings performance, reliability, and governance to the data lake, which houses NLP data as well, first in the bronze layer as the source of truth and then further down the medallion architecture. The silver zone is where all structured variables are extracted from unstructured text and augmented with existing structured sets for enriching downstream analysis, as shown in the following pipeline:

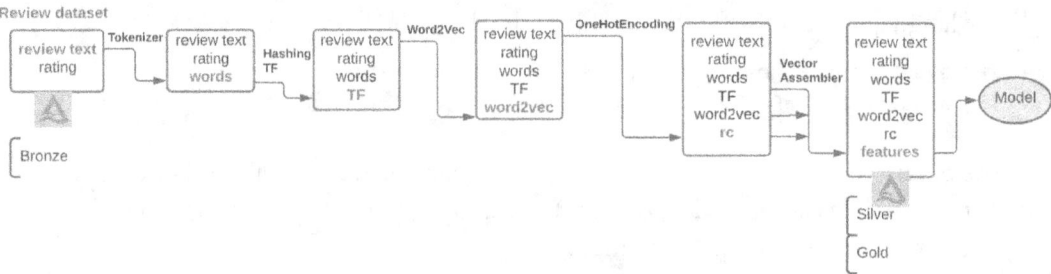

Figure 10.3 – Voice of the customer NLP pipeline

Voice of the customer use cases of NLP refer to data collected at various touchpoints with the customer and are especially important because a concern can trigger a larger flame leading to regulatory and legal battles that risk the reputation of the concerned organization. So, it is common to use a classification model to sift through thousands of customer interactions and label and bin them appropriately so that the most pressing ones get priority and are handled swiftly.

There are some distinct stages of an NLP pipeline:

- Tokenization refers to breaking sentences down into tokens, that is, words.

- Hashing term frequency refers to counting the frequency of terms.

- Word2Vec is an unsupervised technique for converting terms to vectors.

- The vector assembler is the feature creation phase where the data is made ready to pass to a model.

Another example of where expediting the analysis of NLP data is critical is in the health and life sciences space where mining clinical notes (written and audio) and PDF pathology reports in a timely manner can help save lives, as shown in the following figure:

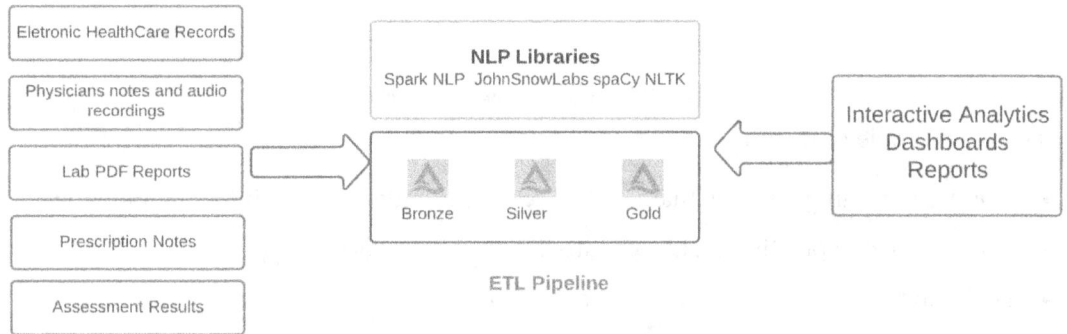

Figure 10.4 – Healthcare NLP pipeline

The various data sources have text and audio content that needs to be handled at scale using several NLP libraries, such as John Snow Labs.

In the next section, we will explore image and video data.

## Image and video data

Image processing was once an esoteric use case, however now it is fairly well understood and there are a lot of models available to train it either from scratch or to use it as a pretrained model. Object detection and object identification are common use cases. The next step is handling video data.

Video files can be considered as a series of image frames and an additional preprocessing step will sample the frames and treat them as image data with a common identifier for the video that can be analyzed in the ML pipeline, as shown in the following diagram:

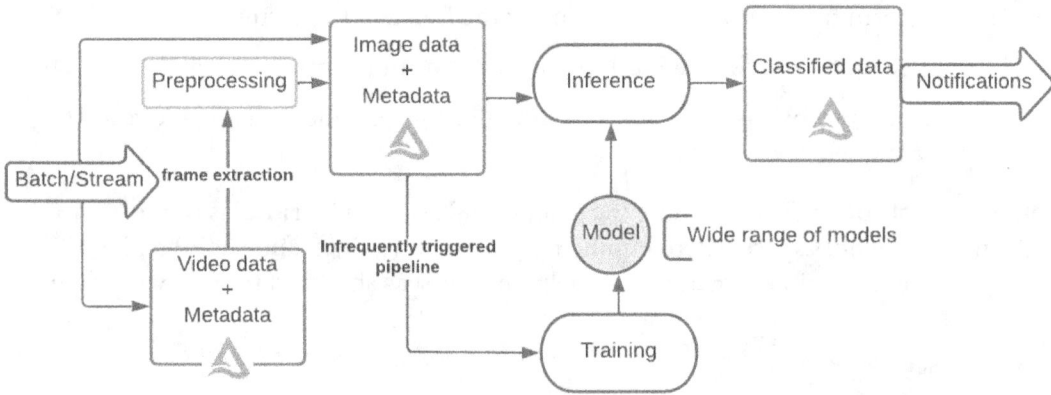

Figure 10.5 – Video/image pipeline

This shows multiple pipelines:

- There is a training pipeline that is called less frequently than the inferring pipeline.

- There is a video pipeline that feeds into an image preprocessing pipeline.

- Data can be fed either in batch or streaming mode.

- The insights generated are persisted in a structured format and additional notifications are triggered based on critical events, such as detecting a poacher in deep forests or detecting the abnormal walking gait of an elderly person and alerting the caregiver to prevent a fall.

Let's break down the video pipeline by processing zones:

- **Bronze**

  - Read raw video files:

```
In: spark.read.format("binaryFile")
Out: UDF explode(split_video_files)
```

- Save them in Delta format:

```
In: spark.read.format("binaryFile")
Out: df.write.format("delta")
```

- **Silver**

  - The individual frames are extracted by splitting video files in memory using libraries such as ffmpeg wrapped in a UDF. This process is parallelized across the various worker nodes from data in Delta tables or views:

```
In: sql("SELECT * FROM videos")
Out: UDF explode(extract_frames)
```

- Persist frames as Delta format along with any additional metadata in the Silver or Gold zone:

```
In: spark.read.format("binaryFile")
Out: df.write.format("delta")
```

There may be additional requirements to incorporate libraries, such as Densepose, Detectron, and OpenPose, to generate additional features to capture poses in a mesh-like frame for analysis of movement and to anonymize the subjects involved.

In the next section, we will see how to marry disparate datasets and look at them holistically.

# Data mashups using Delta

Data mashup refers to combining different datasets to provide a unified data view for analytics. A simple example of this is a BI dashboard combining a consumer's interaction with a brand. The browsing and purchase history are transactional structured data; log data is semi-structured data; the tweets, support cases around product inquiry or complaint; social media posts and comments including text and images are examples of unstructured data that can provide insights into the voice of the customer and user sentiment.

The important elements can be extracted, aggregated, predicted, and brought together with actual transactional data to predict the consumer's next move. The ability to use SQL to query complex aspects of unstructured data and join it with a primary key, such as the customer ID, is very powerful and empowers self-service capabilities. Marketing dollars spent on personalized advertisements and product recommendations can then be purposefully lined up to maximize the probability of a positive impact on the customer journey.

There are two approaches to mashing data. In the first scenario, data from different sources is physically brought together, `blended`, and `harmonized` to make it appear contiguous and unified. In the second scenario, a federated query engine has the ability to query data from its original source and unify it as part of a processing query. The following subsections will illustrate the differences.

## Data blending

It is important to call out that `joining` data is different from the concept of data `blending` in that joins are usually made with tables from the same data source and blending is the process of combining the tables typically across datasets.

Blending conjures mental images of a juicing blender that combines multiple vegetables and juices to give you a single glass of packed nutrients that may not have been possible from a single ingredient. It provides new perspectives that may have otherwise gone unnoticed as missed opportunities. It allows for better efficiency in breaking data silos and rapid analysis. Different file formats, such as CSV, JSON, and Avro, can be blended into a single performant Delta format for downstream consumption by AI and BI tools or multiple Delta sources can be combined to provide a more potent view of the data. It may also be necessary to point out that data integration is more ETL like further upstream putting data into a single data store, whereas blending is further downstream, typically in the reporting layer, such as Tableau, Data Studios among others, may not be necessarily persisted.

## Data harmonization

Data harmonization refers to building a single view of the truth by combining and unifying different data sources, taking into account the disparity in their formats, fields, dimensions, and availability in different contexts. It sounds similar to blending but is more focused on the quality and utility aspects. It interprets existing characteristics in data and lineage to suggest quality improvements, thereby ensuring reliable data is used for insight generation. It applies to different points of the data refinement pipeline, starting with data collection and consent to harmonize views for data consumption. As data volumes grow and the need to share the data grows as well, it becomes especially important in the context of building social and economic data models. The following diagram captures the iterative nature of the process, which follows four main steps, data capture, data normalization, data analysis, and eventually data reconciliation:

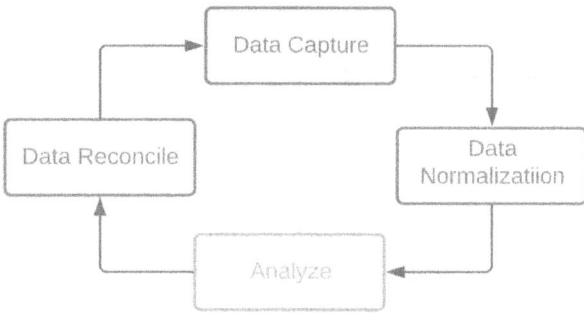

Figure 10.6 – Data harmonization

Let's look at the importance of data harmonization in the context of standardizing clinical data collected at different locations, sent to different labs, preprocessed on differently calibrated machines, and shared with researchers all over the world for use in different research focus areas. Researchers strive to harmonize or standardize their clinical data collected in different labs using differently calibrated machines so that it can be queried in a unified format. The same data point may come through multiple sources and it is important to spot it as a duplicate. This is especially important for data such as a passport application, surgical procedure approval, and an online order. In other words, data homogeneity is necessary to be able to combine data and it can be done by adequately using technology and governance processes.

There are five prescribed steps to get data harmonization:

1. Select the dataset to identify the relevant data sources and the collection process.
2. Standardize data to normalize, correct, and impute columns to improve data quality and produce a clean, consistent dataset.
3. Check the data quality to ensure an acceptable level of integrity and validity over time.
4. Map columns to identify variables for harmonization from multiple sources as they are rarely uniform.
5. Process data to convert it into a common format so that everyone in the organization has the same view of the data and is presented with a subset of data that they have privileges for to suit each LOB or department's needs.

What if we do not want to combine and blend data prematurely? In the next section, on federated query, we will look into how we can query data in place.

# Federated query

Federated query is a concept where the query engine takes care of pulling data from different disparate domain-specific data stores and provides a unified analytic view of the data. It is aligned with the data mesh architecture and there are three main problem areas it attempts to solve, namely the following:

- Avoiding a monolithic centralized data store strategy by allowing domain-specific data teams to have autonomous setups, possibly by business unit. The core data is retained within and the ancillary data needed to augment analytics can be pulled from other data domain products.

- Avoiding generic central policies from being a scale bottleneck for domain-specific processing requirements. Individual data teams have the domain know-how to bring in efficiencies over general policies and recommendations, thereby facilitating agility.

- There are policies around data locality, such as GDPR and CCPA compliance, which is yet another reason to leave certain datasets isolated. The mesh architecture provides connectivity to individual data products.

Data accessibility, availability, and interoperability without copying large volumes of data are the main requirements. You could use distinct Spark DataFrames to collate the data or use a federated query engine, such as Presto, PolyBase, or QueryGrid, to employ predicate pushdown to ensure the filtering happens in each domain and only the relevant datasets get pulled in a single query.

# Facilitating data sharing with Delta

JDBC/ODBC connections or HTTP connections via REST APIs are good for sharing modest data but may become a bottleneck for larger datasets. Consider the scenario of sharing curated data with external vendors or partners. There are some firms whose business model is centered around data sharing, such as S&P, Bloomberg, FactSet, Nasdaq, and SafeGraph. They aim to be the source of truth for financial datasets, which every other financial institution will be interested in consuming for downstream analysis and to augment their own datasets. Wouldn't it be nice not to have to copy the data multiple times?

It is best to use cloud storage access directly to avoid unnecessary platform-related bottlenecks. That is what Delta sharing attempts to do – provide an open standard to securely and seamlessly share large volumes of data in Parquet/Delta with a wide variety of consumers and an easy way to govern and audit. Consumers can be from pandas, Apache Spark, Rust, and other systems without having to first deploy a specific compute pattern, thereby providing a lot of flexibility.

There are already several clients or data consumers that are open source (such as Spark, pandas, Presto, Trino, Hive, and Rust) and commercial clients across BI (for example, Power BI, Tableau, Looker, and Qlik), analytics (such as Databricks, Azure, GCP, Dremio, AtScale, and Starburst), and data governance (for example, Privacera, Immuta, Collibra, and Alation), and there are more coming.

## Setting up Delta sharing

Let's look at how Delta sharing is set up:

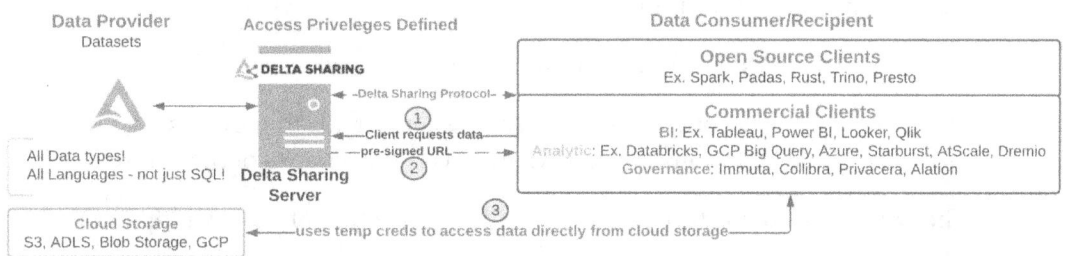

Figure 10.7 – Delta sharing setup

The steps include the following:

1. **The data provider**

   I.   Starts with an existing Delta table with data in Parquet format.

   II.  Defines the access privileges in the Delta sharing server. There is a reference server for the open source community and managed platforms such as Databricks provide it out of the box so that their user base can avoid having to set up and maintain the Delta sharing server themselves. (`https://github.com/delta-io/delta-sharing`)

2. **The data recipient**

   I.   Requests access to a dataset or a subset of it.

   II.  Server sends a short-lived presigned URL to the client.

   III. Clients on a wide range of platforms go straight to the cloud storage.

Now that we've seen how Delta sharing works, let's look at all the other benefits that it brings.

## Benefits of Delta sharing

The four overarching goals of Delta sharing design include the following:

- Avoid copying existing data to avoid delays and errors while accessing *fresh* data.
- Allow for a wide range of consumers that support open data formats, such as Parquet, so that there are no vendor lock-in requirements.
- Built-in security, governance, and auditing to avoid embarrassing breaches so that consumers only get access to the subset of data that they have privileges for.
- Ability to scale the datasets because we are in big data land!

Now that we've seen how Delta sharing is set up, it is fair to say that all four goals are met and there are even more benefits of this simple elegant design, namely the following:

- Open protocol and easy to set up for both parties:
  - Recipient can be provisioned with a few simple commands to get access to a full dataset, just a data partition, a filtered view, or certain versions of the dataset:
    - The first step is to create a `share`, which is a collection of data. Multiple tables can be added to the share.
    - The next step is to add a recipient, which results in an `activation link` that will be used by the recipient to download temporary credentials, which is a `<.share>` file:

```sql
%sql
CREATE SHARE <share_name>;
ALTER SHARE <share_name> ADD TABLE <table_name> ;
DESCRIBE SHARE <share_name>;
CREATE RECIPIENT  <recipient_name>;
GRANT SELECT ON SHARE <share_name> to RECIPIENT <recipient_name>;
```

- Clients even outside the organization and on a completely different platform just make a single call to get the short-lived signed URL to securely access data directly from cloud storage:

```
import delta_sharing

#list the tables added to the share
delta_sharing.SharingClient(<path to delta share credential file>).list_all_tables

# read as either a spark or pandas dataframe or directly from a favorite BI tool
df_spark=  delta_sharing.load_as_spark(<path to delta share credential file>)

df_pandas =  delta_sharing.load_as_pandas(<path to delta share credential file>)
```

- Transfer is fast, cheap, reliable, and read in parallel. The scale of the table is not a limitation.

- Sharing is not limited to tables; the data asset could be a stream, view, arbitrary files, and even ML artifacts.

- The fact that the presigned URL is short lived ensures that it doesn't fall into the wrong hands and get misused as it is time bound.

- A lot of connectors have already been written by the community, leading vendors, and data providers, and more are yet to come, making Delta sharing not just open, but truly ubiquitous.

Data, like oil, is meant to flow. Governance is hard to enforce on data lakes where it is done typically at a file level. Traditionally, data warehouses managed it at the table level but this is fraught with proprietary dos and don'ts, which leads to vendor lock-in, as well as the fact that it does not scale to very large datasets. Delta sharing is the open, reliable protocol that is the answer to those hard problems.

# Data clean room

Collecting and curating data is an involved process that requires substantial time and money. If done right, the datasets have high value and can be used as a monetizing opportunity. A data clean room is a walled garden where organizations share aggregated data that is free of personal markers with some security checks and balances. The aggregated data never leaves the clean room. Vendors and partners come into this space and bring in their datasets to marry them with these aggregated sets and compare across platform offerings to understand user attribution. Companies cannot access any data that is not theirs.

For example, advertisers can check the efficacy of their marketing dollars and where they get the best returns. It is hard as data scales and also gets more expensive. This is where Delta sharing shines as scale is built in and is an inherent consideration as compared to other offerings on the market.

Data clean rooms are an infrastructure concept that has existed for a while, with Google announcing it first in 2017 (Ads Data Hub), followed by Facebook and Amazon. These have been interspersed with stricter adoption of GDPR and CCPA user privacy laws. They are important because they allow marketers to harness the power of the combined data set while adhering to privacy regulations.

The following diagram illustrates how customer data is collected in various applications by providers, anonymized to hide personally identifiable information, and brought into a secure data clean room where consumer agencies are granted access to see their data and optionally join with other datasets to analyze campaign efficiency:

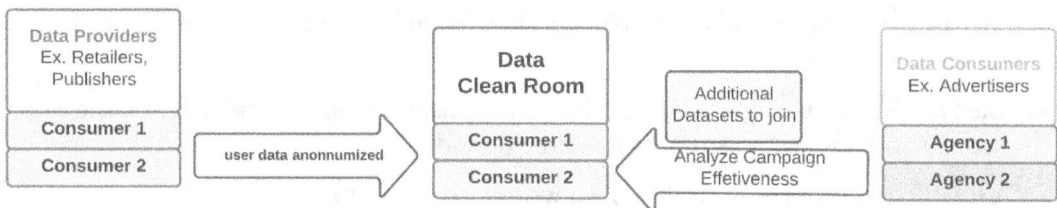

Figure 10.8 – Data clean room

Secure data sharing with well-guarded and governed data access controls from a governance perspective provides an opportunity to utilize the wealth of customer data generated without compromising privacy laws. It is a win-win situation as the consumers learn valuable insights and the providers use it as a monetizing opportunity.

# Summary

DaaS is a natural extension to **software as a service** (**SaaS**), where a data product is made available to a qualified user on demand in a self-service style. Organizations value quality data insights and hence are willing to trade them for tangible benefits. Data piracy and leaks are challenges that need to be considered thoroughly as one significant breach could potentially finish an organization. This is not restricted to just structured data and applies to any data type and at any scale, which makes it an exceedingly hard problem.

In this chapter, we looked at Delta's capabilities around handling more complex unstructured data and how to integrate it with the rest of the structured data so that there is not only additional value from it, but it also democratizes data by allowing ubiquitous access via SQL or other languages. We looked at the importance of data harmonization in the context of governance and a single unified view of normalized, quality data for organization-wide consumption.

Data mashups are relevant in the context of data visualization and reporting of disparate but connected datasets. Data is the lifeblood of an organization and is heavily used in-house, but there is a growing appetite to allow it to flow beyond the borders of the organization where it was created. Data exchanges allow seamless integration with third-party data and are on the rise as a next step in the data economy. We looked at Delta sharing in the context of secure, open data sharing as there is a growing need among data providers and consumers to curate once and consume multiple times in order to reduce the time to analytic consumption.

In the next chapter, we will look at operationalizing data and ML pipelines.

# Section 3 – Operationalizing and Productionalizing Delta Pipelines

A lot of ML projects fail to see the light of production. Beyond a POC, there are several considerations to take a pipeline to production and ensure it is resilient and runs 24*7, adapting to changes in data patterns and business needs to continuously provide value. Cost and performance are important considerations as data scales. Automation is mandatory to continuously refine and evolve a pipeline. ML models once unleashed are living assets and their life cycle management is yet another layer of the underlying data pipeline that needs to be monitored and managed to ensure they produce the expected ROI for the business.

This section includes the following chapters:

- *Chapter 11, Operationalizing Data and ML Pipelines*
- *Chapter 12, Optimizing Cost and Performance with Delta*
- *Chapter 13, Managing Your Data Journey*

# 11
# Operationalizing Data and ML Pipelines

*"We are what we repeatedly do. Excellence, then, is not an act, but a habit."*

*– Aristotle*

In the previous chapters, we saw how Delta helps to democratize data products and services and facilitates data sharing within the organization and externally with vendors and partners. Creating a **Proof of Concept** (**POC**) happy path to prove what is feasible is a far cry from taking a workload to production. Stakeholders and consumers get upset when their reports are not available in a timely manner and, over time, lose confidence in the team's ability to deliver on their promises. This affects the profitability metrics of the business.

In this chapter, we will look into the aspects of DevOps that harden a pipeline so that it stands the test of time and people do not end up spending more time and effort maintaining the system than was required to create it. There is a checklist of items to consider, such as SLAs, high availability, data quality, and automation, that every production pipeline should consider as part of its design and implementation. In particular, we will look at the following topics:

- Why operationalize?
- Understanding and monitoring SLAs

- Scaling and high availability
- Planning for disaster recovery
- Guaranteeing data quality
- Automation of CI/CD pipelines
- Data as code – an intelligent pipeline

Let's look at how Delta can help harden and operationalize data and ML pipelines.

# Technical requirements

To follow the instructions of this chapter, make sure you have the code and instructions as detailed in this GitHub location: `https://github.com/PacktPublishing/Simplifying-Data-Engineering-and-Analytics-with-Delta/tree/main/Chapter11`.

`https://delta.io/roadmap/` is the list of features coming to open source Delta in the near future. This chapter refers to some of them, including Delta clone.

We discuss a Databricks-specific feature called **Delta Live Table** (**DLT**) to give an example of what to aspire for in an intelligent pipeline.

Let's get started!

# Why operationalize?

Consistently bringing data in a timely manner to the right stakeholders is what data/analytics operationalization is all about. It looks deceptively simple, but according to Gartner, only about 10-15% of AI/ML projects truly succeed, and the main reasons are a lack of scale, a lack of trust, and a lack of governance, meaning that not all the compliance boxes are checked to deliver the project within the window of opportunity. The key areas that need attention to enable this include getting complete datasets, including unstructured data, which is the hardest to tame, accelerating the development process by improving means of collaboration between data personas and having a well-defined governance and deployment framework.

By now, the medallion architecture should be a familiar architecture blueprint construct. It is to be noted that in the real world, several producers, several pipelines, and several consumers criss-cross. Each pipeline transforms and wrangles data based on the requirements of the business use case. Ideally, that is what the data personas truly care about. Unfortunately, this is a small part of their overall responsibility. It is a given that no matter how well a pipeline has been tested, it will encounter some errors, therefore gracefully recovering is an important consideration. It is further exacerbated by pipeline dependencies. We need to know when an error occurs. Sometimes the pipeline runs as usual, but the processed data has degraded in quality. That is as good as a failure and needs to be monitored and reported.

Other things that need to be taken care of but can sometimes be overlooked include the partition strategy, dependency management with other tasks and pipelines, checkpointing the state of the pipeline and providing for retries to account for transient glitches in the ecosystem, version control, quality checks, ensuring good cataloging practices for data discoverability, data governance (which includes access privileges and audit trails), and promoting workloads to different environments by creating the underlying infrastructure, among others.

To reiterate, maintaining data quality and reliability at scale day in and day out is a non-trivial task, and even when all the ancillary factors have been considered, the process can be fairly complex. So, anything that can be done to templatize and create reusable components that can be tested and automated goes a long way towards shortening the end-to-end life cycle from development to production. Data is being generated at an enormous pace and it is not surprising that streaming pipelines are becoming the norm to ensure better end-to-end latency.

# Understanding and monitoring SLAs

A **Service Level Agreement (SLA)** is part of an explicit or implicit contract attesting to certain service quality metrics and expectations. Violation of some of these could result in penalties, fines, and loss of reputation. There is usually a cost and service quality tradeoff. So, it is important to articulate the SLA requirements of each use case and describe how it will be measured and tracked so that there is no ambiguity of whether it was honored or violated. There should also be clear guidance on how SLA violations are reported and the obligations and consequences on behalf of the service provider to remedy or compensate for the breach.

There are several types of SLA, and common ones include metrics for system availability, system response time, customer satisfaction as measured by **Net Promoter Score (NPS)**, support tickets raised over a period of time, defect/error tickets and the response time to address them, and security incidents. It is important to distinguish SLAs from KPIs. SLAs refer to compliance expectations, whereas business KPIs are metric indicators.

Latency, cost, and accuracy are key requirements. Let's look at some examples of SLAs:

- The end-to-end data availability from ingestion to consumption is 5 minutes.
- The availability of the service is 99.99% every year.
- The durability of the data is 99.9999% across multiple **availability zones (AZs)**.
- Data should be encrypted both at rest and in motion.
- The maximum query response time of any analytic query on the system is 3 minutes.

Architecting the compute horsepower of clusters requires selecting the right attributes and is determined by the use case being considered. You can think of common SLAs in the context of workload types:

- **Batch**: The volume of data that can be handled by a batch
- **Streaming**: The **Transactions/Events per second (TPS)**
- **BI**: The end-to-end latency and concurrency that can be tolerated
- **High Availability**: The response time to a request
- **Disaster Recovery**: **Recovery Time Objective (RTO)** and **Recovery Point Objective (RPO)**

Now let's look at examples of KPIs:

- Employee turnover should be less than 10%.
- Year-over-year growth should be 15%.
- Incoming ticket volume should be 25 per day.
- NPS score should be greater than 8.

Now that we have established the need for KPIs and SLAs, let's look at the need for constant monitoring to ensure compliance or to detect violations. Data observability refers to understanding the health of data pipelines and goes beyond monitoring and alerting. It is where metrics are continuously computed and logs are scanned to understand the additional context of relevant events, such as who accessed what dataset. Atomic commits to the delta log help with this. Transaction log commits include adding/removing a file and changes to schema.

Let's look at what needs observing:

- **Schema** changes that can disrupt operational pipelines and affect key business processes.

- **Lineage** helps to identify, isolate, and determine the root cause of broken pipelines.

- **Volume** changes are usually within an acceptable threshold, and violating this threshold at either end suggests the unavailability of a data source or the addition of new ones.

- **Statistical** characteristics of a data distribution should remain similar, and wide variations indicate changes to data collection and the interpretation of base assumptions on the data.

- **Freshness** of data refers to the fact that data flows through the systems and the fresher the data, the faster the insight generation and relevance to business.

Let's see how Delta helps address these observability needs:

- Delta supports disciplined schema evolution, allowing innocuous changes early in the pipeline by allowing mergeSchema and locking it down towards the latter stages to not surprise the consumers or violate the established data contract with them.

- Delta Time Travel provides built-in lineage and visibility to operations for future audits. This helps with reproducibility as well. However, transformations across Delta tables need to be tracked explicitly.

- Volume counts on every batch or time period can be set up with thresholds to alert on anomalies. This is not Delta-specific, but it auto-captures a lot of this in the transaction log.

- Data profiling using Spark profiling or pandas profiling libraries can add value when the statistical characteristics of data change significantly.

- Delta's support for unifying batch and streaming with tunable ingestion frequency offers an easy means to control data freshness based on SLA needs and cost concerns, meaning lower latency requires continuous compute and hence higher costs.

In the next section, we will look at the availability of pipelines to ingest new data and serve queries for stored data.

# Scaling and high availability

**Scalability** refers to the elasticity of compute resources, meaning adding more compute capacity as data volume increases to support a heavier workload. It is sometimes necessary to scale down resources that aren't in use to save compute costs. Scaling can be of two types: vertical or horizontal. **Vertical scaling** refers to replacing existing node types with bigger instance types. This is not sustainable after a point because there is an upper bound on the largest possible instance. **Horizontal scaling** refers to the addition of more worker nodes of the same type and is truly infinitely scalable. Each serves different scenarios. If the largest partition is no longer divisible, we benefit from a bigger node type. However, the advantage is that some of the nodes can be turned off when there is low data volume. This is an infrastructure and architecture capability and not directly related to Delta.

**High availability** (**HA**) refers to the system uptime over a period of time to service requests. There are several underlying nuances. For example, a pipeline may encounter an out-of-memory error or a human coding bug and become unavailable. For a real-time use case, this will immediately have an impact on the SLAs. A particular instance type may become unavailable, such as a spot instance or a GPU instance. Some GPU workloads may not be able to function, but spot instances could be replaced by on-demand instances for a higher price. Spark's inherent distributed architecture and cloud architecture ensures that a worker node is replaced by another in the event of a node's unavailability. You will be more vulnerable to meta store limits than object store rate limits. Delta will scale beyond meta store limits since the transaction log manages the creation of the logical table instead of the meta store. As a result, Delta tables are not dependent on a meta store because the details are in the transaction log, which lies alongside data and is easy to replicate elsewhere. However, the use of a meta store can still facilitate data discovery.

AZ failures can be addressed more easily than total region failures. It is important to note that HA is in the context of localized errors and load scenarios where **disaster recovery** (**DR**) is a more widespread catastrophic failure requiring a shift in infrastructure to ensure business continuity. In the next section, we will look at different options for business continuity by allowing a DR strategy to another region.

# Planning for DR

Planning for DR requires a balance of cost and time needed for a business to recover from an outage. The shorter the time expectation, the more expensive the DR solution.

It is important to understand two key SLAs for the business use case:

- **Recovery Time Objective (RTO)** refers to the duration in which a business is mandated to recover from an outage. For example, if RTO is 1 hour and it is 30 minutes since the outage, then we have 30 more minutes to recover and bring the operations back online without violating the RTO stipulations.

- **Recovery Point Objective (RPO)** refers to the maximum time period of a disruption after which the loss of data collection and processing will exceed the business's agreed-upon threshold. For example, if backup was done in the last hour and the defined RPO is 2 hours, we still have an hour to recover from the disruption to the business.

In the next section, we will see how to use these values of RTO and RPO to plan for DR.

## How to decide on the correct DR strategy

Before designing a DR strategy, it is important to understand these two metrics. DR strategies can be categorized in the following four ways, with increasing cost and decreasing RTO/RPO values, as shown in the following diagram:

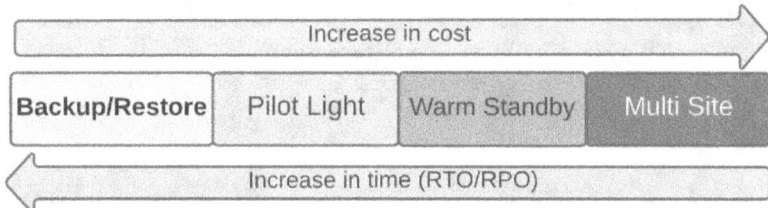

Figure 11.1 – DR strategies

There are four main types of DR strategy. From left to right, the cost of the strategy increases, and the time involved in recovery decreases. So, depending on the SLAs and budget at hand, the right DR strategy should be chosen, and they are as follows:

- Backup/Restore:

  - Data is replicated.

  - Infrastructure scripts are versioned and available to both primary and secondary sites. The repository could be Git (which also could be a single point of failure) or just cloud storage with replication to a different region.

- Pilot Light:

  - A second deployment site exists so that in the event of failure, both code and scripts would need to be deployed to restart operations there.

- Warm Standby (Active/Passive):

  - Not only does a secondary site exist, but all the artifacts and code are also up to date, they're just not processing data yet.

- Multi-Site (Active/Active):

  - Both sites are completely primed and ready and a load balancer shunts traffic to each site so that in the event of a failure to one region, the full load falls on the other.

Let's see what role Delta plays in facilitating a DR strategy.

## How Delta helps with DR

These are the main ingredients to required establish a DR site in another region:

- Data (including metadata)
- Scripts/code/libraries needed to run your pipelines
- Code for your infrastructure setup/deployment

We covered Delta deep clones in *Chapter 6, Solving Common Data Pattern Scenarios with Delta*, and now we will see how handy it is to replicate data to the secondary region. There are other ways of copying data using cloud-specific geo-replication strategies. However, they have deficiencies that could lead to additional data copies and inconsistent state. Also, they are restricted to being only one way. Delta deep clones are not only incremental but also two-way, meaning you can fall back to the secondary region in the event of a failure in the primary site and also go back to the primary site once it is reinstated. Note that the metadata replication is also necessary and can be done by periodic exports and imports, and it is preferable to rely on an external meta store. This is required only if the downstream user/workload accesses the table via the meta store; otherwise, the transaction log copied as part of a deep clone operation is sufficient.

The following gives an example of how to create the clone and query both tables to validate data reconciliation:

```sql
%sql
CREATE OR REPLACE TABLE <delta_clone_table> DEEP CLONE <delta_table>;

-- Original view of data
SELECT * FROM <delta_table>

-- Clone view of data
SELECT * FROM <delta_clone_table>
```

In the RDBMS world, replication is facilitated using the commit logs, and the same is true with Delta: the transaction logs are used to perform the synchronization between the primary and secondary sites. It is important to emphasize that this incremental clone is synchronous in nature, making sure that data order is maintained even in a distributed architecture setup. Changes include not just additional records but updates and deletes to existing data. The following diagram demonstrates the steps involved in incremental data transfer between the primary and secondary sites, the failover to secondary in the event of a disaster in the primary, and the subsequent fallback to primary as order is restored there:

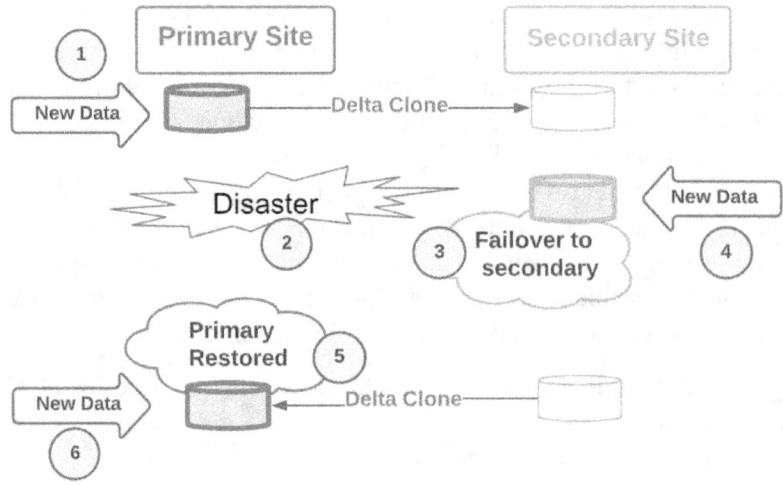

Figure 11.2 – DR failover and fallback

Let's look at the six steps illustrated in the diagram:

1.  Updates are made to the data at the primary site. The data is incrementally copied/cloned to the secondary site.

2.  Disaster strikes the primary site.

3.  The pipeline processing fails over to secondary site, which takes over the pipeline and business operations.

4.  Changes are made to data in the secondary site.

5.  The primary site is restored.

6.  The pipeline can now fall back to the primary site using a new clone table and it can be renamed. Changes are now made to the primary site and we can continue to clone to the secondary site as usual.

DR should be an important consideration in your business continuity plan. In the next section, we will look at quality concerns of the data, which, if not handled adequately, render the data untrustworthy and hence useless.

# Guaranteeing data quality

We've examined the medallion architecture blueprint, where raw data is brought as is into bronze, refined in silver, and aggregated in gold. As data moves from left to right in the pipeline it is getting transformed and refined in quality by the application of business rules (how to normalize data and impute missing values, among other things), and this curated data is more valuable than the original data. But what if the transformations, instead of refining and increasing the quality of the data, actually have bugs and can occasionally cause damage? We need a way to monitor the quality and ensure it is maintained over time, and if for some reason it degrades, we need to be notified. If there is an occasional fix by updating the data, it is a needle in a haystack scenario, but nevertheless, it needs to be accommodated easily.

Delta's ACID transaction support ensures that in the event of a failure, no data is committed, ensuring that consumers never see bad or partial data. This would be a nightmare to rewind from in a distributed architecture with multiple nodes. The same is true of Delta's support for schema evolution, where a pipeline engineer controls whether mergeSchema should be allowed or not depending on where the data is in the pipeline. In the event of an inadvertent human error, Delta's time travel capability allows data to be rolled back to a previous time or version. Similarly, Delta's support for fine-grained updates and deletes allows data fixes to be applied to a few data points with relative ease. Storing metadata in transaction details along with the data in object storage helps to recreate an old dataset.

However, there are a few more features that need attention and need to be coded either in individual pipelines or as part of a framework for greater configuration and reusability:

- Providing rules for acceptable values of certain fields. Examples include ranges, enumerations, relational expressions, and regex patterns.

- Providing actions (rules) on how violations should be handled. Some options include reducing quality metrics, dropping the data, quarantining it, or halting the pipeline.

It is necessary to build a data quality audit framework on data in the lake to achieve the following:

- Define quality rules as policies so that they can be adopted, standardized, and changed over time. These become contracts that are enforced on each data ingest.

- At-a-glance visibility of the quality metrics of a dataset is necessary for folks downstream to make informed decisions on whether to use the dataset or not.

- Actionable triggers on violation, such as alerting or quarantining the data points. The baseline metrics are monitored for drift, and any anomaly is logged for audit purposes and reported to invite intervention and fixing.

- Because there are stages of data dependencies, integration with data lineage is an important consideration. One bad transformation could significantly lower the quality of a curated dataset that was otherwise pristine.

The following diagram extends the medallion blueprint with additional validation checks by data engineering folks and business folks so that the datasets that have been explicitly validated have additional marker flags to indicate quality metric values. Note the process of releasing the data to downstream users could be a design tradeoff. In some cases, it is made available and businesses can consume it at their own risk. In other cases, it may be required to clone the data to a separate consumption layer table to keep it completely pristine and separate. This does add to the end-to-end delay and is a bottleneck in self-service activities, so it should be considered carefully.

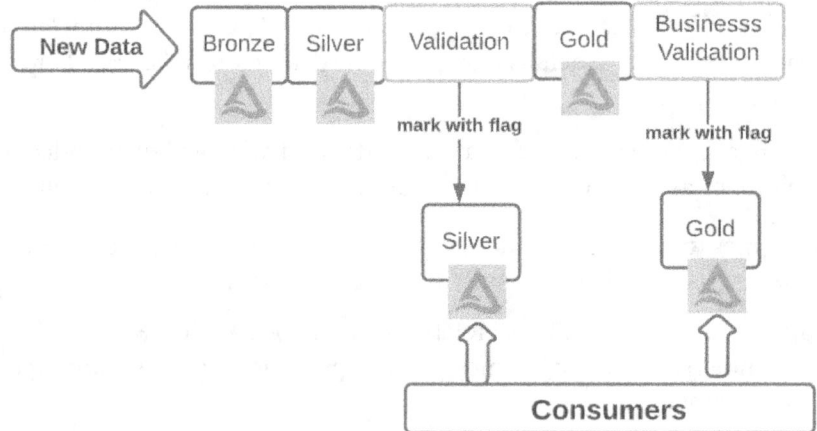

Figure 11.3 – Quality framework

Maintaining data quality is a shared responsibility, and all data personas, including business stakeholders who consume the data, should collaborate to help with the standardization and reporting of anomalies. This requires several building blocks, including a configuration store, a rule engine, anomaly detection and reporting, and a graphical interface for visual display. In a managed platform such as Databricks, this is solved by the DLT feature, where quality is built into the table definition and is not left to the discretion of a data persona to implement as an afterthought. Maintaining this as a separate step or process often results in the data and quality checks not being in sync.

In the next section, we will look at testing and automation that facilitates building and maintaining pipelines at scale.

## Automation of CI/CD pipelines

POC and Pilot code to prove out an end-to-end path does not get sanctioned for production as is. Typically, it makes its way through dev, stage, and prod environments, where it gets tested and scrutinized. A data product may involve different data teams and different departments to come together and test the data product holistically. An ML cycle has a few additional steps around ML artifact testing to ensure that insights are not only generated, but also valid and relevant. So, **Continuous Training** (**CT**) and **Continuous Monitoring** (**CM**) are additional steps in the pipeline. Last but not least, data has to be versioned because outcomes need to be compared with expected results, sometimes within an acceptable threshold.

Automation takes a little time to build, but it saves a lot more time and grief in the long run. So, investing in testing frameworks and automation around CI/CD pipelines is a task that is worth investing in. **Continuous Integration** (**CI**) is the process of fostering innovation and improvement by having a process to continuously make changes to the code base and submit it after due review and testing. **Continuous Delivery** or **Deployment** (**CD**) refers to the deployment of the new approved changes in an automated fashion to higher environments.

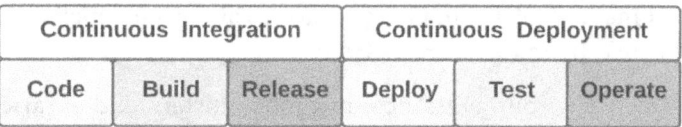

Figure 11.4 – CI versus CD

With CI, regressions can be caught early on, meeting the release deadline gets more predictable, and operational continuity is not a nightmare with constant escalations and triages. The built-in Spark test suite is designed to test all parts of Spark.

With CD, software complexity is tamed and release cycles can be shorter, meaning customers get fixes and feature updates faster as the feedback loop is shorter.

# Code under version control

Often, people think of their pipeline code as the only artifact to protect against loss. However, all aspects of recreating the environment, including infrastructure deployment scripts, configurations, compute specifications, and job and workflow orchestration specifications, all need to be treated as code. So, best practice is to preserve all of this under version control. The Git repository could be cloud-based or hosted on-premises. At times, even a Git repository could have an outage, and this could lead to some work not being saved or even risk your DR strategy. Some firms use a replication process using cloud storage where Git artifacts are copied to different S3 buckets, for instance.

Organizing code into libraries with modular functions facilitates testing and helps to manage complexity, otherwise the pipeline is fragile and breaks easily. Special attention should be made to not expose secrets and credentials and to grant access to repositories judiciously to avoid intellectual property exposure apart from unwanted hacking.

# Infrastructure as Code (IaC)

Leveraging REST APIs and **command-line interfaces** (**CLIs**) helps maintain IaC. Sometimes, cloud provider offerings such as Cloud Formation in AWS may help to abstract away the infrastructure pieces quickly. However, given that all large enterprises are planning to use a multi-cloud strategy to reduce dependencies on a single cloud provider and hedge their risks, using a cloud-agnostic scripting framework such as Terraform may be a better idea. There may be some effort required to initially learn how to use the tool, but it goes a long way to completely automate the release process. Scripts can be versioned and made available in Git repositories alongside pipeline code, so it is no surprise that this is called **Infrastructure as Code**.

The creation of complete environments, including jobs, clusters, and libraries, can be handled programmatically, and it becomes part of the CI/CD pipeline. Terraform in particular has a declarative syntax, keeps track of infrastructure state, handles dependencies between objects, and can be used to not only create but update and delete environments along with their associated objects.

# Unit and integration testing

The natural order of testing is unit test, integration test, and end-to-end tests. Costs rise exponentially as bugs are found later in the development cycle. So, pushing all testing to the end is not a good idea. Removing the human in the loop and automating this process makes this process repeatable and sustainable.

unittest is a popular unit testing framework from Python, and we will use it to demonstrate how to organize tests into suites to incrementally add tests and increase code coverage (reference: https://docs.python.org/3/library/unittest.html).

Sometimes the functions already exist and you can reuse them; other times you may need to create them explicitly and add checks using assert statements that validate results to expected outcomes. The following snippet shows a SimpleTest class with two test functions:

```python
import unittest

class SimpleTest(unittest.TestCase):
    def test_1(self):
        df = spark.sql(f"select * from …")
        self.assertEqual(df.count(), 300000)

    def test_2(self):
        predicted = get_data_prediction()
        self.assertEqual(predicted, 42)
```

Now the test suite class can be generated by adding the individual test functions from before:

```python
def generate_test_class_suite():
    suite = unittest.TestSuite()
    suite.addTest(SimpleTest('test_1'))
    suite.addTest(SimpleTest('test_2'))
    return suite
```

It is also possible to discover the functions by listing everything and adding them or selecting specific ones:

```
def generate_function_suite(suite = None):
    if suite is None:
        suite = unittest.TestSuite()
    suite.addTest(unittest.FunctionTestCase('test_1'))
    suite.addTest(unittest.FunctionTestCase('test_2'))


    return suite
```

This is how the test suite is orchestrated using a test runner:

```
runner = unittest.TextTestRunner()
runner.run(suite)
```

Not all batch operations are supported in streaming, so although it is tempting to just test the logic using batch dataframes, there may be some nuances of source/sink that may be overlooked. It is recommended to use the `StreamTest` harness (Scala only) using memory sink for this. Similar to `assert`, there is `CheckAnswer` and `ExpectFailure`:

```
testStream(stream, OutputMode.Update)(AddData(inputData, …),
CheckAnswer(("a" -> 1))
```

Of course, end-to-end integration testing should be used if unit testing is not adequate or is hard to set up. If all goes well, stress testing should be done to understand load and rate limits so that the clusters can be sized adequately. Monitoring is the other side of the coin. You can get the last progress of a streaming query, for example, (`streamQuery.lastProgress`) to understand input versus processing rate, current processed offsets, and state metrics. Setting up a `StreamingQueryListener` with functions such as `onQueryStart`, `onQueryEnd`, and `onQueryProgress` allows us to get the progress asynchronously.

The following diagram captures the entire process:

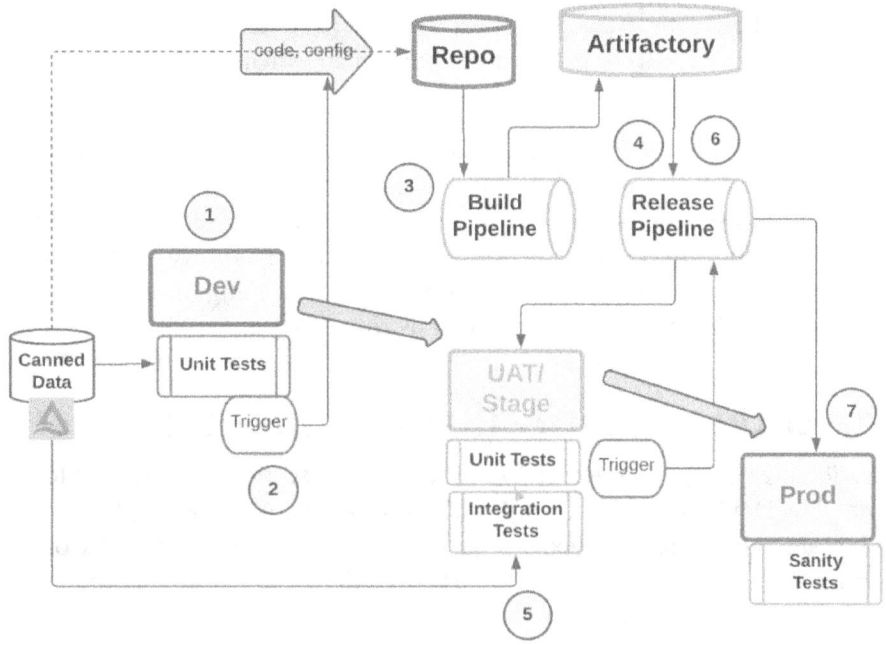

Figure 11.5 – CI/CD to promote pipeline across environments

Let's look at the different flows enumerated in this diagram:

1.  The developer successfully executes unit tests in the development environment.
2.  The developer then pushes code base to a repository.
3.  This triggers a build pipeline. Artifacts are created and pushed to UAT/Stage for unit testing. On successful completion, artifacts are optionally pushed to an artifactory.
4.  A release pipeline is triggered that pushes artifacts from the artifactory to a UAT environment.
5.  Successful execution of integration tests triggers another release pipeline.
6.  The release pipeline pushes the artifacts into the production environment
7.  In the production environment, sanity tests are run and if errors are encountered, then a rollback to a previous version may be required.

Shallow clones (aside from CI/CD use) are also very useful for POCs for optimization/enhancements on already operationalized pipelines with virtually no risk. They solve the problem of "how do I get data to test X?" where it's been pretty common to either read a prod table (which is not ideal) or generate dummy data (which is also not ideal and is time-consuming).

These steps are a coarse outline of the expected workflow. Delta's versioning capability helps to record the version of the dataset to test along with configuration details for subsequent testing across different environments. Schema changes and rollback scenarios can be handled gracefully. In the next section, we will envision a future of intelligent pipelines that abstract away a lot of the admin functionality, leaving the developer to focus on the data transformation aspects as demanded by the use case.

# Data as code – An intelligent pipeline

All the operationalizing aspects referred to in the previous sections would have to be explicitly coded by DevOps, MLOps, and DataOps personas. A managed platform such as Databricks has abstracted the complexity of all these features as part of its DLT offering. The culmination of all these features out of the box gives rise to intelligent pipelines. There is a shift from a procedural to a declarative definition of a pipeline where, as an end user, you specify the "what" aspects of the data transformations, delegating the "how" aspects to the underlying platform. This is especially useful for simplifying the ETL development and go to production process when pipelines need to be democratized across multiple use cases for large, fast-moving data volumes such as IoT sensor data.

These are the key differentiators:

- The ability to understand the dependencies of the transformations to generate the underlying DAG and schedule them to run autonomously. The DAG captures the data lineage of data transformations across tables. In previous chapters, we saw how Delta maintains the version history and lineage of all transformations done to a single table. This takes a step further and captures the lineage of data as it touches multiple tables during the transformations.

- The ability to use configuration to move from batch to streaming and vice versa without requiring any code changes.

- The ability for configuration to move to a different environment, such as dev to prod, without requiring recoding and retesting.

- Simplifying the **Change Data Capture** (**CDC**) process, where the merge operations are further simplified and optimized with sophisticated add-ons such as handling out-of-order events in a distributed setup.

- The use of features such as autoloader (on Databricks) allows the detection of new files, handling schema evolution, and placing unparseable data in rescue data columns for further investigation, thereby ensuring data is never lost.

The following example shows creation of a DLT on Databricks:

```
%sql
CREATE STREAMING LIVE VIEW <bronze_delta_table> AS
SELECT * FROM cloud_files('/data', 'csv');
```

- Delegating the choice of compute, along with autoscaling decisions for the workload at hand, to the underlying platform. Judicious autoscaling, especially for streaming workloads, by monitoring both the backlog metrics and the cluster utilization is especially challenging.

- Delegating administrative aspects of the pipeline run, such as retry on failure, and of the pipeline maintenance tasks, such as running optimize and vacuum, to the platform because it knows the load conditions and can determine the best time to schedule admin tasks to be least disruptive.

- Documentation is built into the data transformation and can be exposed in catalogs for users to discover.

- The ability to build quality constraints and expectations into the pipeline so they are integral and tightly coupled instead of being separate independent jobs that will eventually not be able to keep up with changes made to the pipeline. These constraints should have actionable consequences. For example, if a violation occurs, we can decide how to handle it. Should it be allowed to pass with reduced quality, should it be quarantined or dropped, or should the whole pipeline be halted?

The following example shows constraint definitions for a DLT:

```
%sql
CREATE STREAMING LIVE TABLE <bronze_delta_table> AS
( CONSTRAINT valid_account_open_dt
    EXPECT (dt is not null and (close_dt > open_dt)
) ON VIOLATION DROP ROW
COMMENT "Bronze table with valid account ids" ;

SELECT * FROM <raw zone table> ;
```

- The observability of the pipeline operations, along with the quality metrics, so that people can understand the health of the pipeline and people downstream can decide whether the data quality is reliable enough to be used as input for their processes.

DLT is regarded as the new gold standard for Delta pipelines as it makes the data ingestion pipelines smarter and more robust. Infrastructure is managed in an optimized manner, thereby relieving the user from the tooling aspects so they can focus on the business transformations on the data to reap value. The establishment of quality constraints makes stakeholders trust their data and generate other insights from it with higher confidence.

## Summary

Organizations rely on good data to be delivered in a timely manner to make better business decisions. Every use case has SLAs and metrics that need to be honored. So, operationalizing a pipeline starts with an understanding of both the functional and non-functional business requirements so that people are not surprised that it either does not comply with expectations or is too expensive. With thousands of data pipelines spanning multiple lines of businesses and their inter-dependencies, it is a non-trivial task to ensure they all run successfully and the data they produce is complete and reliable.

In this chapter, we examined the various aspects to be considered when building reliable and robust pipelines and ensuring they continue to run in spite of environmental issues to ensure business continuity. In addition, we explored the need for lineage tracking, observability, and appropriate alerting so everyone is on the same page and can make decisions on when to consume them for their insight generation and reporting needs. The transactional nature of Delta, coupled with features such as Delta clone, makes it easy to build quality frameworks on top of the data lake to ensure all data meets the expected quality standards.

In the next chapter, we will look at performance-related features of Delta and tuning opportunities to continuously refine and improve your Delta pipeline to extract the maximum value from your investment.

# 12
# Optimizing Cost and Performance with Delta

*"You have to perform at a consistently higher level than others. That's the mark of a true professional."*

– *Joe Paterno, Football Coach*

In the previous chapter, we saw how Delta helps with the hardening of data pipelines and making them production worthy. Now that the pipeline is put into action, crunching massive datasets every day, we want that pipeline to be as lean and mean as possible, to extract every ounce of performance from it to make the most of the infrastructure investment while serving the needs of the business users. In addition, the numbers around the end-to-end SLA requirements of pipelines need to be shaved. Both cost and speed are the driving requirements, which is why the key metric is *price performance* as both aspects need to be optimized.

In this chapter, we will explore the ways that data engineers can keep up with the performance demands at an optimum price point. Data patterns change over time, so tuning activities have to be continuously performed to keep up with them. In particular, we will look at the following topics:

- Improving performance with common strategies
- Optimizing with Delta
- Is cost always inversely proportional to performance?
- Best practices for managing performance

Let's look at how Delta helps to extract every ounce of performance from your pipeline.

# Technical requirements

To follow along with this chapter, make sure you have the code and instructions as detailed on GitHub here:

```
https://github.com/PacktPublishing/Simplifying-Data-
Engineering-and-Analytics-with-Delta/tree/main/Chapter12
```

`https://github.com/delta-io/delta/issues/920` is the proposed roadmap for select edge feature migration from Databricks to open source Delta primarily for performance enhancements.

Let's get started!

# Improving performance with common strategies

The performance of a pipeline refers to how quickly the data load can be processed. Throughput is defined as the volume of data that can be processed. In a big data system, both are important scalable metrics. Let's look at ways to improve performance:

- **Increase the level of parallelism**: The ability to break a large chunk into smaller independent chunks that can be executed in parallel.
- **Better code**: Efficient algorithms and code help to crunch through the same business transformations faster.

- **Workflow that captures task dependencies**: Not all tasks can run independently; there are inherent dependencies between tasks and pipelining or orchestration refers to chaining these dependencies as DAGs, where the inherent lineage determines which ones can run simultaneously and which ones need to wait until all the dependent stages have completed successfully. Even better would be the option to share compute for some of these tasks and the ability to repair and run only the affected part of the pipeline in case of a localized failure. Relying on external time schedulers to stitch the task invariably fails as an unexpected spike in data volume or temporary glitch in resource availability can cause the timelines to slide and tasks start to stampede on one another.

Distributed architectures such as Spark can scale both vertically and horizontally to finish the job faster. Vertical refers to larger worker nodes and horizontal refers to a greater number of workers. There is an upper limit to vertical scaling bound by the largest available server node. Although there is no limit to horizontal scaling, the mere addition of parallel worker nodes may not yield better performance because the workload may not be designed with the right partition strategy to take advantage of the load distribution. Spark's cost-based optimizer improves query plans by using table- and column-level statistics. ANALYZE TABLE commands are used to collect these stats.

In the Hadoop world, a partition is the lowest common denominator. So, if a record needs to be modified or deleted, the whole partition is dropped and recreated, which is a very expensive operation. A partition consists of several files and so dealing with file-level granularity allows for faster operation, which is how Delta achieves its fine-grained operations as well as the associated performance benefits. Optimum file size then becomes an important consideration. Too small or too large both have negative ramifications on query performance. Known queries are those whose access pattern is well known and they contribute to dashboards and reports. Providing configuration knobs for easy change to tweak these by use case is an important consideration that Delta provides.

Understanding common patterns and optimizing for those is imperative. This is where optimizations around data layout and the proximity or locality of relevant data that is queried together comes in handy. Interactive exploratory queries are also important but the user is usually a little more patient and can tolerate longer delays as they usually scan much bigger data volumes. Usually, it is a good idea to bucket them into small, medium, and large size and provide SLA expectations on each category.

There are some standard things to look out for when trying to tune a workload:

- What is the real bottleneck – CPU, memory, or I/O?
- What horsepower/node type is right for the job?
- Does it need horizontal or vertical scaling?
- What is the range of auto-scaling that is prudent?
- Will it benefit from GPUs?
- Is there caching?

A pipeline involves several distinct stages and it is important to ensure that they are all performant. These stages include the following:

- **ETL aspects**
  - Ingesting data.
  - Processing and transformations to model the data for the use case at hand in the lake.
  - Enriching data for relevant stakeholders.
  - The primary persona involved here is that of a data engineer.

- **Consumption aspects**
  - Getting data ready for consumption by BI and AI personas.
  - Ingestion strategies may not match the consumption patterns and this may require different partitioning strategies, file layout, and so on. For example, while designing a pipeline for ingestion, you may be concerned about how quickly to bring in the data, so your partitioning could be by processing date, whereas a consumption pattern may care about the event date. So, it is important to understand what patterns are most common and cater to those first.
  - The primary personas involved here are that of a business analyst or an ML practitioner.

Tuning activities in all these stages are additive and help get the insights in the hands of the stakeholders faster, reducing the analysis time and execution of the actionable insights.

The following diagram makes the distinction between the performance optimization activities in the ETL stage and the additional considerations for the consumption stage. In addition, there are administrative activities that need to be periodically applied to the data to tune file size and generate statistics or get rid of old versions of the data that are no longer necessary. In the next section, we will look at these Delta-specific operations, namely the following:

- OPTIMIZE helps with file compaction to avoid the small file problem.

- ANALYZE generates additional runtime statistics metadata that is stored in the meta store that helps during operations such as **Adaptive Query Execution (AQE)** to modify and refine the query plan on the fly.

- VACUUM removes data files (not log files) that are no longer referenced by the Delta table and are older than the configured retention threshold, which by default is 7 days.

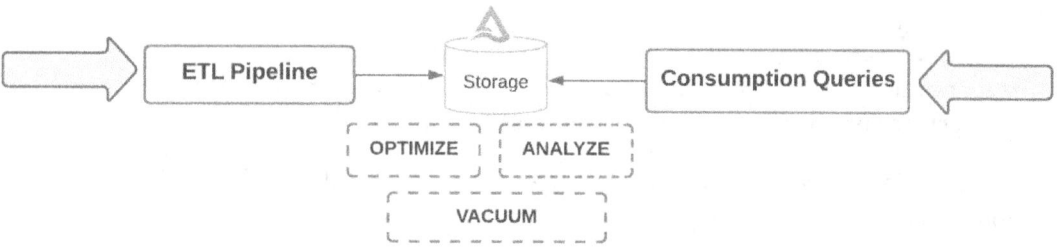

Figure 12.1 – Performance tuning for ingestion versus consumption patterns

In the next section, we will look at where to look for slow queries and what to look for in order to begin the optimization process incrementally.

## Where to look and what to look for

The Spark UI offers several views to the underlying query plan so that you can spot trouble-zone bottlenecks and address them appropriately by providing more compute, memory, or I/O. The explain command can be used to review the plan and verify that generated statistics are used to optimize queries. Big data pipelines can suffer from the following syndromes:

- **Spill**

    - When the dataset is too large to fit in memory, it will spill to disk, and disk operations are more expensive, which is why spills should be avoided.

- **Shuffle**

  - When the data operation requires the movement of data across workers, the internode communication could slow down operations, which is why shuffles should be reduced. Operations such as `groupBy`, `orderBy`, and `sort` will result in a shuffle.

- **Skew/stragglers**

  - When the mapping of tasks to cores is not well laid out, it results in the last few tasks running on a few worker nodes while the rest of the nodes are passively waiting, burning compute but not actively participating in the computation.

- **Slow functions**

  - These are usually unoptimized user-defined functions that need to be reviewed and optimized. The thread dump of the executors will reveal slow functions.

- **Small files**

  - Continuous ingestion of microbatches of data leads to small files, which affects query performance and hence needs coalescing and compaction.

In the next section, we will see what options Delta provides to address some of these common concerns.

# Optimizing with Delta

Delta's support for ACID transactions and quality guarantees helps ensure data reliability, thereby reducing superfluous validation steps and shortening the end-to-end time. This involves less downtime and triage cycles. Delta's support of fine-grained updates, deletes, and merges applies at a file level instead of to the entire partition, leading to less data manipulation and faster operations. This also leads to fewer compute resources, leading to cost savings.

## Changing the data layout in storage

Optimizing the layout of the data can help speed up query performance, and there are two ways to do so, namely the following:

- **Compaction, also known as bin-packing**

  - Here, lots of smaller files are combined into fewer large ones.

- Depending on how many files are involved, this can be an expensive operation and it is a good idea to run it either during off-peak hours or on a separate cluster from the main pipeline to avoid unnecessary delays to the main job.

- The operation is idempotent, so running it more frequently does not involve additional compute if there no new data has arrived since.

- The `optimize` operation can be done on a table, on a file path, or on a smaller subset of the table by specifying a `where` clause.

- The default target file size is 1 GB. However, you can alter that by specifying the value in bytes of two parameters: `optimize.minFileSize` and `optimize.maxFileSize`.

- The default value of the maximum number of threads that run in parallel to run the `optimize` command can be increased by tweaking `optimize.maxThreads`.

- **Data skipping**

  - This feature is enabled by default and results in metrics collected on the configured (`delta.dataSkippingNumIndexedCols`) first set of columns of the table, such as min/max values of a column, which helps with faster queries. Long value columns, such as strings, should be placed later to avoid expensive stats being collected for them. The recommendation is to structure tables to move all numerical keys and high-cardinality query predicates to the left and long strings and dates to the right to take advantage of the automatic statistics collection. These stats are kept in the Delta transaction logs. This can be set as a table property as well.

In the next section, we'll examine optimizations offered by managed platforms.

# Other platform optimizations

The next few optimizations are provided on the managed platform offering on Databricks; however, some of these make their way into open source as well:

- **ZOrder,** also known as multidimensional clustering, is a file layout technique to colocate related information together because the access patterns use those fields heavily in their `where` clauses:

- To determine which columns to use for z-order, look for high-cardinality fields that frequently appear in the WHERE clause or are used in the JOIN criteria of queries. Typically, these are identifier fields, such as user ID, IP address, product ID, email address, and phone number. In conjunction with the file-skipping feature, this offers a better selection and orders of magnitude can be shaved off a query on large datasets as the file statistics help to hone in on just the relevant ones. This is useful in almost all large-scale queries, such as cybersecurity analysis involving several terabytes of data to detect **intrusion detection patterns (IDPs)** with stringent SLAs.

- For example, date can be a partition field and peril_code can be a zorder field for an insurance claim dataset. 3 or 4 is a good number of columns to consider for z-order if needed, more than that will cause them to be less effective. It is not an idempotent operation; however, if no new data has arrived since the last run, it is a no-op operation. .

- The following is an example of applying zorder on select columns and potentially select partitions of a Delta table. It can be used in conjunction with partitions, where partition columns are chosen for categorical fields and date fields with lower cardinality so that a substantial number of files (at least 1 GB) sits within a partition to make its creation worthwhile:

```
OPTIMIZE <delta_table> [WHERE <partition_filter>]
ZORDER BY (<column>[, …])
```

- **Auto-optimize**: This is particularly relevant for streaming workloads and is similar to optimize in the sense that it collates smaller files, except that it happens in flight before it hits the disk for all write operations. There is an adaptive shuffle that happens just before the write operation, so fewer files need to be written. It can be set either at the table level or for all tables, as shown:

```
ALTER TABLE <delta_table> SET TBLPROPERTIES 'delta.autoOptimize.optimizeWrite' = 'true';
SET spark.databricks.delta.properties.defaults.autoOptimize.optimizeWrite = true;
```

In addition, if the requests are coming in at a high throughput, it is necessary to prevent hotspots. For example, in the context of an object store such as S3, a simple way to do this is to use random prefixes:

```
ALTER TABLE <delta_table> SET TBLPROPERTIES 'delta.randomizeFilePrefixes' = 'true'
```

- **Dynamic file pruning** (**DFP**): DFP is a data-skipping technique to improve queries with selective joins on non-partition columns on tables in Delta Lake. Static partition pruning is off filters, whereas dynamic is off joins and is a file pruning strategy that kicks in at runtime as opposed to compile-time, hence the name dynamic.

  An example of this is a join where one side of the join can use pushdown predicate details to return fewer joins if the other side has far fewer rows and can communicate the same so that the total amount of data brought together for the join is less:

  ```
  SET spark.databricks.optimizer.dynamicFilePruning = true

  -- minimum table size on the probe side of the join required to trigger (DFP)
  -- default is 10G
  SET spark.databricks.optimizer.deltaTableSizeThreshold = <> bytes

  -- number of files of the Delta table on the probe side of the join required to trigger dynamic file pruning
  SET spark.databricks.optimizer.deltaTableFilesThreshold = <N>
  ```

- **Bloom filter index**: Each data file can have a single associated Bloom filter index that can help you answer that data is definitely *not in* the file but there could be cases of false positive. This index file is first checked and the file is read *only if* the index indicates that the file *might* match a data filter. It is an uncompressed Parquet file that contains a single row and is stored in the _delta_index subdirectory relative to the data file:

  ```
  -- Enable the Bloom filter index capability
  SET spark.databricks.io.skipping.bloomFilter.enabled = true;

  CREATE BLOOMFILTER INDEX
  ON TABLE <delta_table>
  FOR COLUMNS(sha OPTIONS (fpp=0.1, numItems=50000000);
  ```

- **Optimum file size**: Too-small or too-large files are both bad for performance. If there are too many small files, more time is spent in file I/O operations instead of actually reading data. If the files are too large, they may need to be split, or if they are not splitable, then the workers are not utilized well. It is important to set an optimum file size depending on your use case, as shown:

  ```
  SET spark.databricks.delta.optimize.maxFileSize = 1610612736;

  OPTIMIZE <delta_table>
  ZORDER BY <fields>
  ```

- **Low Shuffle Merge** (**LSM**): Merge operations can be very expensive on account of the number of rows they touch. LSM improves performance by processing unmodified rows in a separate processing mode, instead of processing them together with the modified rows. So, the amount of shuffled data is reduced significantly, leading to improved performance:

```
SET spark.databricks.delta.merge.enableLowShuffle = true
```

- **Optimize joins**: AQE is a Spark feature and should be enabled by default as it does a lot of the optimizations on behalf of the user. There might be some scenarios where you need to specify hints that both range and join optimizations benefit from. The bin size is a numeric tuning parameter that splits the values domain of the range condition into multiple bins of equal size. For example, with a bin size of 10, the optimization splits the domain into bins that are intervals of length 10.

  Data skew is a condition in which a table's data is unevenly distributed among partitions in the cluster and can severely downgrade the performance of queries, especially those with joins. A skew hint must contain at least the name of the relation with skew. A relation is a table, view, or subquery. All joins with this relation then use skew join optimization.

  The bin size can be set explicitly or as part of the hint, as shown:

```
SET spark.databricks.optimizer.rangeJoin.binSize=5

SELECT /*+ RANGE_JOIN(points, 10) */ *
FROM points JOIN ranges ON points.p >= ranges.start AND points.p < ranges.end;
```

- **Delta caching**: It creates copies of the cloud storage data locally on an SSD disk on worker nodes so that subsequent reads are faster and is supported for Parquet/Delta formats. Unlike Spark Cache, this process is fully transparent and does not require any action, and memory is not taken away from other operations within Spark. Eviction is automatically done in **least recently used** (**LRU**) fashion or on any file change manually when restarting a cluster:

```
SET spark.databricks.io.cache.enabled = true;
SET spark.databricks.io.cache.maxDiskUsage = <> ;
SET spark.databricks.io.cache.maxMetaDataCache = <> ;
SET spark.databricks.io.cache.compression.enabled = true;
```

The following diagram shows the position of Delta cache in an application data stack:

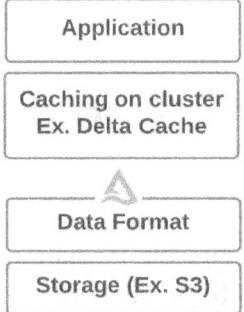

Figure 12.2 – Delta cache

Although the cache need not be explicitly invalidated or loaded, to warm up the cache in advance, the CACHE SELECT command can be used. If the existing cached entries have to be refreshed, the REFRESH TABLE statement can be used, which is lazily evaluated.

- **Optimize data transformations**: Higher-order functions for operating on arrays and complex nested data type transformations can be very expensive. Using Spark's built-in functions and primitives is recommended in array manipulation operations.

In the next section, we will look at ways to make this process as automated as possible; REST APIs are the way to go about it.

## Automation

For most read/write operations, the regular Spark APIs can be used. However, there are some operations that are specific to Delta that require DeltaTable APIs (refer to https://docs.delta.io/latest/api/python/index.html):

```python
from delta.tables import *

deltaTable = DeltaTable.forPath(spark, "/path/to/delta_table")
deltaTable.optimize()
deltaTable.vacuum()
```

The `DeltaTable` class is the main one for interacting with the underlying data. All the data operations that we've talked about, such as `update`, `delete`, and `merge`, can be executed using methods on this class. You can use it to convert Parquet to Delta or generate a manifest file so that processing engines other than Spark can read Delta.

# Is cost always inversely proportional to performance?

Typically, higher performance is associated with higher costs. Spark provides options for tunable performance and cost. At a high level, it is a given that if your end-to-end latency is stringent or low, then your cost will be higher.

But using Delta to unify all your workloads on a single platform brings efficiencies of scale through automation and standardization, leading to cost reductions by reducing the number of hops and processing steps, which translates to a reduction in compute power. Also, when your queries run faster on the same hardware, you pay for a shorter duration of your running cloud computing cost. So yes, it is possible to improve performance and still contain the cost. SLA requirements are not compromised. Instead, superior architecture options are available, such as the unification of batch and streaming workloads, handling both schema enforcement alongside schema evolution, and the ability to handle unstructured data right alongside traditional structured and semi-structured data. The simplification of pipelines, along with increased reliability and quality, leads to fewer outages and triages/fixes freeing up people from support tasks to work primarily on improving use case effectiveness.

It is important that when you are evaluating product architectures, you look for cost performance metrics. The cost of individual services can be misleading, so the cost of the entire pipeline should be considered for an effective comparison. Duration, failure rates, the ability to support current and future use cases, including critical workloads and their SLAs, and the engineering effort to build and maintain the pipelines should all be considered in the mix. At the end of the day, the only thing that matters is what `value` is delivered from your data investment. Also, remember that not every pipeline needs to be as fast as possible; it should match the consumption pattern. If there is no one to consume data at the other end, then it is wasteful to refresh it too frequently.

The areas to consider include the following:

- Ease of setup
- Ease of use
- Ease of extensibility and integrations

- Ease of migration to another location
- Ease of moving from development mode to production
- Ease of moving from low-frequency to high-frequency scheduling
- Ability to support both data and ML pipelines along with REST APIs for CI/CD automation and observability

The metrics to consider can be summarized as follows:

- Total cost of running the entire pipeline
- Time duration from end to end to get the data ready and the access patterns to consume it
- Productivity gains for the engineering and business teams
- Extensibility of the architecture to future-proof for unknown use cases
- Complexity involved in the creation and maintenance
- Portability of the solution so you do not sign up for an expensive vendor lock-in situation

In the next section, we will summarize some of the best practices to get the best performance.

# Best practices for managing performance

Managing cost and performance is a continuous activity. Sometimes, they can inversely affect each other, and other times, they go hand in hand. Once optimized, a workload pattern can change and need a different set of tweaks. That said, managed platforms, such as Databricks, are getting better at analyzing workloads and suggesting optimizations or directly applying them, thereby relieving the data engineer from these responsibilities. But there is still a long way to go to reach complete auto-pilot. We covered a lot of different techniques to tune your workloads; partition pruning and I/O pruning are the main ones:

- **Partition pruning**: It is file-based by having directories for each partition value. On-disk, it will look like `<partition_key>=<partition_value>` and a set of associated Parquet data files. If the amount of data pulled from executors back to the driver is large, use `spark.driver.maxResultSize` to increase it. It may also suffer from too many files, which could slow down the writing operation. Use `optimize` and `auto-optimize` to optimize writes and auto-compaction. You should examine the Delta log table.

- **I/O pruning**: Data skipping and ZOrdering help with the better management of the granularity of file size and avoiding skews in data file size.

- **Others**: Choose cluster configuration judiciously. Add a TTL policy or disable cloud storage versioning to avoid a lot of storage and set `spark.sql.shuffle.partitions` to the number of executor cores. Use Delta caching to locally cache data and avoid going to cloud storage. This is especially useful in search applications where there is high query reuse or small changes in filter conditions.

Here are some common best practices to follow:

- Employ the right compute type.

- Optimize based on the need of the workload under consideration - storage versus compute versus memory.

- Once a family type is chosen, start with the lowest configuration and understand bottlenecks before throwing larger node types.

- Depending on SLAs, choose a reserved instance for the driver and a combination of reserved and spot instances for workers as spot instances can greatly reduce cost.

- Benchmark to size the cluster but allow for auto-scaling (both up and down).

- In most scenarios, these setting should be applied:

  - Turn on AQE (`spark.sql.adaptive.enabled`).

  - Turn on coalesce partitions (`spark.sql.adaptive.coalescePartitions.enabled`).

  - Turn on skew join (`spark.sql.adaptive.skewJoin.enabled`).

  - Turn on local shuffle reader (`spark.sql.adaptive.localShuffleReader.enabled`).

  - Set the broadcast join threshold (`spark.sql.autoBroadcastJoinThreshold`).

  - Avoid `SortMerge` join (`spark.sql.join.preferSortMergeJoin = false`).

- Build pipelines that capture dependencies of all associated tasks so that downstream tasks are auto-triggered when relevant, dependent upstream tasks are complete.

- Review and refine code so that it is as lean and mean as possible. Parallelize wherever possible. Look for the five Ss in the Spark UI and other monitoring dashboards and address them early on. Set up alerts and notifications when a job takes too long or fails.

- Build all pipelines to not just use Delta but streaming as well – the dial can easily be changed to move from less frequent to more frequent to continuous streaming ingestion mode.

- Plan to execute scheduled maintenance tasks, such as `optimize` and `vacuum`. It's advised to run maintenance on a separate cluster, with possibly different node configuration, and not on the giant cluster you might have just run your data pipeline on

Here are some tips for making sound decisions around your data layout and operations to maintain data hygiene:

- **What partitioning strategy should you use?**

  Which column(s) to use for partitioning is a common design consideration. If the table is small to modest in size, it may be best to not have a partition at all. However, if you do partition, you should take into consideration access patterns and the `where` columns commonly used. For example, date is a common partition strategy. The rule of thumb is to not use a high-cardinality column, such as identifiers. However, if the data in a partition is expected to be about 1 GB, then it is a fair candidate.

- **How do you compact files?**

  Writing data in microbatches over time creates small files that will adversely affect performance. Compacting using `repartition` or `optimize` will help change the data layout and coalesce the data into fewer files. In both cases, you could do it for the entire table, by partition. In the case of repartition, the older files will need to be removed by running a `vacuum` command.

- **How do you replace content or a schema gracefully?**

  There may be instances where business logic changes and you need to replace a Delta table. You may be tempted to delete the directory and recreate it or create a new one. This is not a good idea as it not only is slower to delete files but also runs the risk of data corruption. A better approach is to use the overwrite mode to replace data with the `overwriteSchema` option set to `true` or fix a few values using Delta's inherent capability for fine-grained updates and deletes.

- **What kind of caching should you use?**

  When using Delta, it is advisable to not use Spark caching as any additional data filters will cause it to go back to disk, as will accessing using a different identifier. Delta caching where possible at the VM level should be employed as the dataset itself is stored locally. As the data in cloud storage changes, the cache gets invalidated so you do not have to worry about accessing stale data.

The primary performance levers are listed in the following diagram:

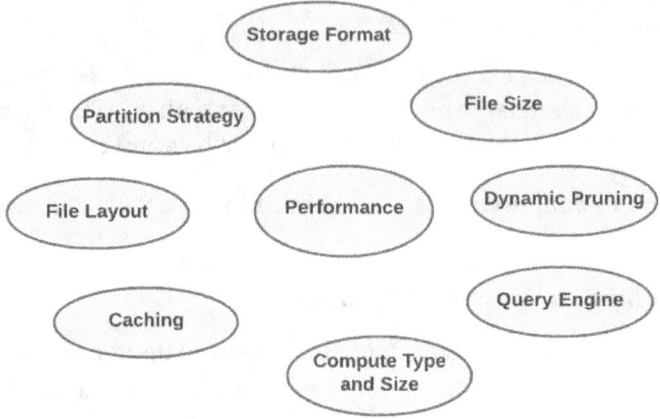

Figure 12.3 – Performance levers

We have explored performance from several dimensions, including cost. The best way is to write queries, examine the query plan, optimize, explore, and learn iteratively.

# Summary

As data grows exponentially over time, query performance is an important ask from all stakeholders. Delta is based on the columnar Parquet format, which is highly compressible, consuming less storage and memory and automatically creating and maintaining indices on data. Data skipping helps with getting faster access to data and is achieved by maintaining file statistics so that only the relevant files are read, avoiding full scans. Delta caching improves the performance of common queries that repeat. `optimize` compacts smaller files and `zorder` colocates relevant details that are usually queried together, leading to fewer file reads.

The Delta architecture pattern has empowered data engineers not only by simplifying a lot of their daily activities but also by also improving the query performance for data analysts who consume the hard work and output produced by these upstream data engineers. In this chapter, we looked at some common techniques to apply to our Delta tables to make them perform better for data analyst queries. In the next chater, we will look at admin-related functionalities, particularly around user management and data access control, so the governance pillar will be the final feather in the cap in the data journey, along with some data migration plays.

# 13
# Managing Your Data Journey

*"You possess all the attributes of a demagogue; a screeching, horrible voice, a perverse, cross-grained nature, and the language of the marketplace. In you, all is united which is needful for governing."*

– Aristophanes, *The Knights*

In the previous chapters, we looked at the roles and responsibilities of the primary data personas, namely data engineers and data scientists, and ML practitioners, business analysts, and DevOps/MLOps personas. One persona that we have not talked about much is that of an **administrator**. They are the gatekeepers that hold the key to deploying infrastructure, enabling users and principals on a platform, setting ground rules on who can do what, being responsible for version upgrades, applying patches, security, and enabling new features, and providing direction for business continuity and disaster recovery, and so on. What will all this look like in a multitenant ecosystem where some lines of business have shared access to data and others don't?

In particular, we will look at the following topics:

- Provisioning a multi-tenant infrastructure
- Democratizing data via policies and processes

- Capacity planning

- Managing and monitoring data

- Data sharing

- Data migration

- **Center of Excellence** (COE) best practices

Let's look at how an administrator goes about planning these various responsibilities. **Delta** may or may not directly relate to each sub-area.

# Provisioning a multi-tenant infrastructure

The administrator is tasked with setting up the infrastructure for the tenants of an environment. One question that often arises is *what should be the optimum balance of collaboration and isolation.* Creating single deployments and putting everyone there could lead to hitting rate limits and is not a sustainable strategy. Since we have the luxury of elasticity of the cloud, we can turn on as many environments as we wish to isolate data and users and provide better blast radius control in case of a security breach. Conversely, creating too many environments leads to harder governance and maintenance challenges, collaboration suffers, and the enablement cycle could be much longer.

Let's examine the various scenarios:

- Separate development, staging, and production environments.

- Disaster recovery requires setting a parallel production environment in a different region.

- Different lines of business units want a separate isolated environment.

- Some lines of business wish to share some resources.

Within an environment, there will be several data personas who need access to compute and storage. Sometimes, these resources can be shared, and at other times, they need to be isolated. Some workloads need regulatory compliance, such as **Payment Card Industry (PCI)**, **Health Insurance Portability and Accountability Act (HIPAA)**, **System and Organization Control (SOC)**, and **General Data Protection Regulation (GDPR)**, which require more attention. Other times, there is a risk of **Intellectual Property (IP)** exposure and data exfiltration that needs additional sensitivity around both the compute and storage handling.

No matter what multi-tenancy strategy you choose, here are some key considerations:

- Data should remain in an open format on cheap, reliable cloud storage so it can be accessed from multiple environments by multiple tools and frameworks. Delta with underlying Parquet is an ideal file format for this.

- The users of an organization are already provisioned in a central **Identity Provider (IDP)**, so it is a good idea to sync them into your environments using something such as **System for Cross-Domain Identity Management (SCIM)**. The advantage is that groups can be synced, so if users get added or removed, that change is automatically picked up.

- These principals (users, groups, and service principals) have entitlements that define their privilege levels of what they can access and what authorizations they have on that data to view, update and delete. Again, these entitlements should be defined once and synced into the various environments so that a user does not inadvertently provide access in one environment, leading to data and IP exfiltration.

- Admins themselves can be of several types. There may be a super admin who can override everything; there may be other account-level admins supporting the super admin and helping to create other environments. Then, each of these environments will be assigned its own admin who has jurisdiction over only a particular environment.

Delta being an open format can be accessed from any environment if the access privileges allow for it. In the next section, we will look at the roles and responsibilities of an admin for a given environment.

# Data democratization via policies and processes

If everything is locked down, then there is no threat of exposure. However, that is not the intended agenda of data organizations. Getting the relevant data in the hands of the right privileged audience helps a company innovate by allowing people to explore and discover new meaningful ways to add business value from the data. IT should not be the bottleneck in the process of data democratization. If new datasets are brought in, IT should not be overwhelmed with tickets from every part of the organization requesting access to them. So, enabling self-service with appropriate security guardrails is an important responsibility of an administrator. This is where policies play an important role in policing an environment, either preventing an unintended situation from taking place or reporting against it by running scans to detect patterns so that bad actors or novices can be corrected in time.

Policies can be of several types; some typical examples include the following:

- Restricting the type and size of compute used in an environment:

  - A 200-node cluster when a 20-node would suffice is wasteful and can rack up a hefty bill. Similarly, using expensive GPU nodes when a workload requires only CPUs can be prevented by providing an allowable list of node types and a maximum size of the cluster.

- Enforcing the use of tags:

  - In a shared environment, usage and billing attribution can be challenging. If every cluster/job is tagged by a team name, then the chargeback model and reporting dashboard can be simpler to interpret. So, enforcing a consistent tagging and naming convention, will help in the long run.

There might be one-off cases, such as a team actually needing 200 nodes where the provided policy templates do not work for a particular team. It is an exception rather than the norm that has not been accounted for because it is very specific and rare. Folks may get blocked and come to a stalemate situation, demanding that policies be loosened to take care of these special situations. You should not buckle and give in to these demands as it will hurt the majority of scenarios. The whole thing was done for better control and governance. Instead, a better way to handle it would be by a process where the affected team gets an exemption approved by a higher-level business executive who justifies that usage, and then a new policy can be created only for that specific group.

A good place to examine how to define policy is to look at a feature offering from a managed platform such as Databricks. So, now, we've established ground rules that others in an environment are going to play by, although rules are made to be broken. In the next section, we will see how to audit the adherence to policies and report against non-compliance.

# Capacity planning

Data volumes are constantly growing. Capacity planning is the art and science of arriving at the right infrastructure that caters to the current and future needs of a business. It has several inputs, including the incoming data volume, the volume of historical data that needs to be retained, the SLAs for end-to-end latency, and the kind of processing and transformations that are done on the data. It is directly linked to your ability to sustain scalable growth at a manageable cost point. We may be tempted to think that leveraging the elasticity properties of cloud infrastructure absolves us from planning around capacity, which is in correct!

So, how do you go about forecasting demand? The simplest way is to use a sliver of data, establish a pilot workstream, take the memory, compute and storage metrics and project it out for the full workload, adding in some buffer for growth and then repeating it for every known use case, while keeping a buffer for unplanned activity. This exercise needs to be done over a 12-month period; in some cases, it may be longer.

The next thing is to determine a percentage of it for lower environments, such as development and staging or a production environment in a different region set up for business continuity or disaster recovery purposes. ML planning is a little harder because it is a true scientific experiment, and data scientists typically run hundreds of architectures before converging on one. In that case, using time and compute to come up with baselines is a more pragmatic approach. The worst thing that can happen is to be surprised by high costs and a use case being turned down or completely shut off because the initial projections were too low.

# Managing and monitoring

Every organization has policies around data access and data use that need to be honored. In addition, there are compliance guidelines in some regulated industries to prove that compliance is honored, using an audit trail of the types of user access and manipulation of the data. Hence, there is a need to be able to set the controls in place, detect whether something has been changed, and provide a transparent audit trail. This includes access to raw data as well as via tables that are an artifact on top of the data.

The metrics collected from these logs need to be compared over a period of time to understand trend lines. Delta's versioning capability comes in handy to monitor not only operations done on a table but metrics logged as well. It would be fair to say that these metrics need more permanence and some date/time stamp would be used to log them.

There are several types of logs in a system. The main ones include the following:

1. **Audit logs**:

   - *Who is doing what* in the various environments – that is, an audit trail of user actions. For example, someone deleted a cluster or edited a pipeline workflow definition. Parsing the events/actions captured in the audit log helps to detect anomalous patterns that may need timely rectification.

   - Administrators, security, and compliance auditors value this information.

2. **Cluster logs**:

- The Spark UI provides a wealth of information around driver/worker logs, `stdout`/`stderr` and `log4j` outputs, and reflects the usage of jobs and the associated data crunching.

- Administrators, data engineers, and ML personas often analyze these logs.

3. **Spark metrics**:

- Metrics around Spark jobs/stages/tasks such as the volume of data read, crunched, and written. Shuffle, spill, and garbage-collect details to help understand performance.

- Typically useful for support and performance engineers looking to debug bottlenecks.

4. **Virtual Memory (VM)** and **system metrics**:

- This is the CPU/memory/storage utilization of the underlying VMs.

- Performance can be monitored via tools such as Gaglia, Datadog, and other monitoring agents.

- Typically useful for support and performance engineers looking to debug bottlenecks.

5. **Custom logging**:

- These are purely purpose-built to provide additional data points, such as detecting data or model drift, to alert data engineers and ML practitioners to reconsider a newer iteration of their baseline logic or model.

- The following diagram provides a framework to monitor and manage the various logs in a multi-tenant environment.

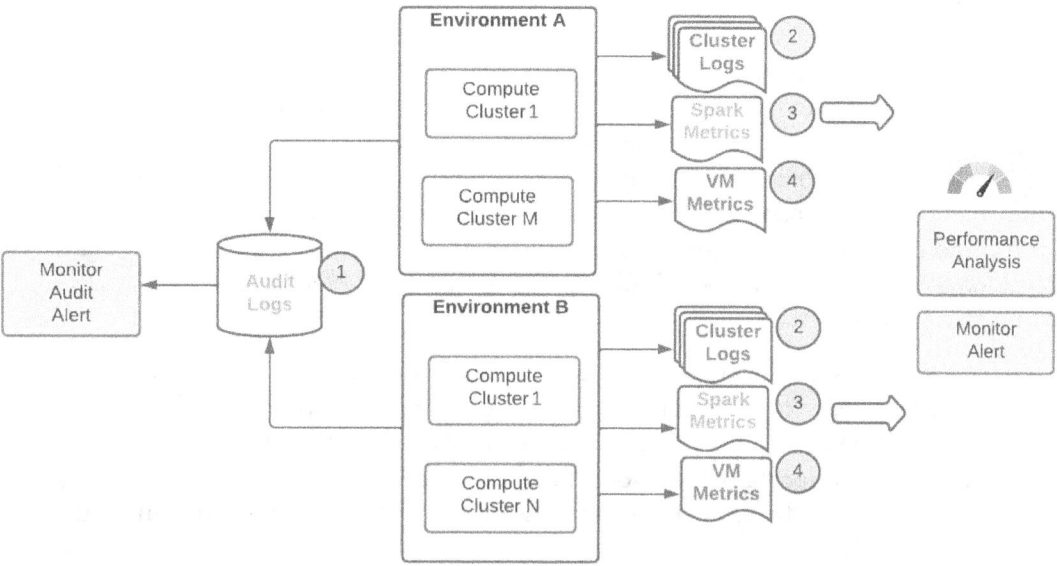

Figure 13.1 – Logging and monitoring your data environments

Certain activities such as SSO provisioning, SCIM integration, and audit logging are provisioned once for all environments. Others such as cluster usage and billing logs are typically collected per environment but may benefit from rolling up as a centralized view. Marrying the audit logs and cluster logs can be very powerful to understand your most expensive workloads and users and help with demand forecasting, as well as tuning activities. This is done by directing them to a centralized cloud storage location, segregated by environment. This may look like a lot of work, but the price of failure and non-compliance is very high. So, all mature organizations should plan to have a solid strategy around logging and monitoring, as they provide a lot of telltale signals about the health of pipelines and how well they are being governed and managed.

# Data sharing

This is a paradox to a lot of collaboration and isolation concepts we reviewed in earlier sections. When groups or lines of business have a lot of data dependencies, they are usually housed together to facilitate better collaboration, and if they do not have any operational dependencies, they can be segregated in their own environments – for example, HR and marketing may be in their own domain meshes. However, what happens if there is a need for them to share some insights? There should be a way to promote it, as it leads to better stakeholder engagement that improves enterprise value. However, all the painful architecting to ensure this accidental exposure does not happen will now have to be reconsidered. That is a lot of unnecessary complexity and re-architecting. Also, data replication to a shared location will lead to the two getting out of sync. Thankfully, Delta sharing comes to the rescue.

A simple, open, and secure way to share data can be achieved through Delta sharing without requiring multiple copies of data and any vendor lock-in propositions. We covered this in a previous chapter so will not go into the mechanics of it again. It suffices to say that the **Delta Sharing** server brokers the exchange between the data provider and the data recipient, and it helps facilitate any BI/AI use case, using any tool on any cloud, including on-premises.

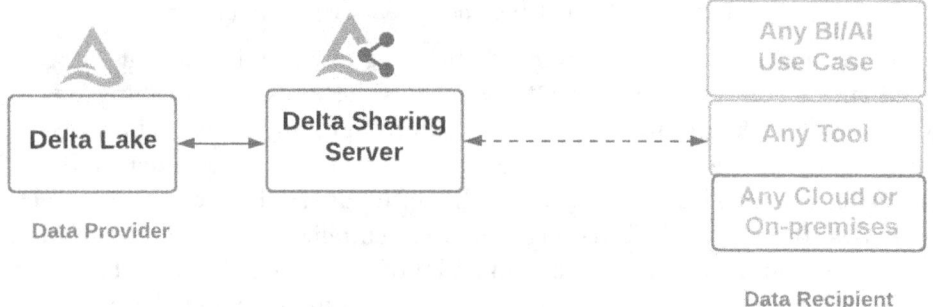

Figure 13.2 – Delta Sharing

But first, let's see how it is different from other similar offerings:

- Secure, cost-effective, and zero compute cost, barring some egress charges when it is cross-region.

- Vendor-agnostic, multi-cloud, and open source.

- Table/partition/DataFrame-level abstraction.

- Scalable with predicate pushdown and object store bandwidth.

- An asset is not only the data but other assets such as dashboards and models. Delta sharing allows for model sharing as well.

Now, let's explore the main use cases benefiting from it:

- Data sharing between **Line of Business (LOBs)** that do not typically have operational dependencies:

  - Expensive complex rearchitecting can be avoided.

  - Moreover, there may be a case where one LOB receives data from two or more siloed LOBs and is responsible for joining and consolidating all the pieces, either by maintaining anonymity or otherwise, and this enables new use cases that would otherwise have been very difficult to pull through.

- Additional opportunities for data monetization:

  - The time and cost it has taken to curate and validate datasets can be useful to other organizations, allowing for an additional revenue stream.

- Data sharing between disparate architectures:

  - All that is needed for Delta sharing is a laptop and the ability to read Parquet.

  - In a multi-cloud deployment, it serves as the glue between the various disparate environments and regions.

In the next section, we will look at another opportunity for large-scale data movement. However, unlike the other scenarios, this is usually a one-time operation.

# Data migration

Technologies are constantly evolving. It is important to choose a platform and architecture that is future-proof and extensible and supports a pluggable paradigm to play nicely with other tools of an ecosystem. So gravitating towards open data formats, open source tooling, and cloud-based architecture with separation of compute and storage, you can dodge the main bullets. There will be a time when this is no longer sustainable and the whole data platform needs a refreshing overhaul. Some examples of this that we've seen in recent years is migration from Hadoop-based systems that are complex and difficult to manage to cloud-native data platforms. The same is true of expensive **data warehousing** solutions such as **Netezza**, **Teradata**, and **Exadata**. Migration projects are expensive, time-consuming, and critical to the overall value of a business and tech investments and need to be planned and executed very carefully.

How will you determine whether to patch an existing system or rebuild it with a newer tech stack? The main driving forces are as follows:

- Expensive **Total Cost of Ownership (TCO)**, which continues to grow as the volume of data grows.

- Inflexible systems that worked well for some use cases but are no longer conducive for newer use cases, either because it is very difficult to build using them or, in some cases, just not possible. For example, leveraging unstructured data for advanced analytic use cases is not something that traditional databases support.

- Vendor lock-in to proprietary formats and technologies where integration with other parts of a data ecosystem is difficult or unsupported.

To mitigate risk and ensure an on-time and on-cost migration, a phased approach is typically followed:

1. A discovery phase where the existing workload is examined using a profiler and consultative approach to benchmark what workloads were running in older environments and bucket them into three areas of straightforward, moderate, and high complexity.

2. The next phase is that of assessing the best fit of people, processes, and technology. This is where tooling and partners are lined up to examine what automation can be achieved by existing tooling and accelerators.

3. This is followed by a migration workshop phase where all concerned stakeholders come together to draw up a master plan and strategy to execute upon. A technology mapping exercise is done at a finer granularity to determine which are the lift and shift jobs and which ones need architecting in the new environment, along with ballpark effort and cost figures.

4. Next comes the pilot phase, where a typical use case is chosen to execute upon. The reference architecture is drawn up and the follow-up roadmap.

5. All this has been a paper exercise. This is an actual pilot implementation phase where the rubber meets the road, and some common pain points are addressed to serve as lesson learning for subsequent iterations. The two environments remain up in parallel, allowing for easy validation.

6. The same is done for all workloads, and rigorous testing and data reconciliation activities are done to prove the migration completion point. After a cutover, the older environment is gradually shut down and permanently retired.

The main technology mappings to consider are as following:

- Data storage
- Metadata storage
- Code migration around compatible libraries and APIs
- Data processing and transformations
- Security
- Orchestration of jobs and workflows

If data is not in Delta format, files can be converted to Delta using one of these options:

- Convert a Parquet table to Delta:

```
CONVERT TO DELTA <parquet table>
```

- Convert files to Delta format and create a table using that data:

```
CONVERT TO DELTA parquet.<`/data-path/`>
CREATE TABLE <delta table> USING DELTA LOCATION <'/data-
path/'>
```

- Convert a non-Parquet format such as ORC to Parquet and then to Delta:

```
CREATE TABLE <parquet table> USING PARQUET OPTIONS (path
<'/data-path/'>)
CONVERT TO DELTA <parquet table>
```

One thing to keep in mind while porting Delta is to avoid bypassing the transaction logs, as they retain the source of truth and can cause inaccuracies. Of course, running vacuum on a Delta table will remove all versions and make it look similar to a Parquet file. You can also generate a manifest file that can be read by other processing engines such as Presto and Athena using the following:

```
GENERATE symlink_format_manifest FOR TABLE <delta table>
```

In the next section, we will look at the need to establish a center of excellence within an enterprise and its roles and responsibilities.

# COE best practices

Establishing an internal steering committee/team as the **Center of Excellence (COE)** for creating advanced analytics is a complex process. Its primary purpose is to provide a blueprint to onboard data teams, and enable them with technical and operational practices, support for handling issues and tickets, and executive alignment to ensure that technical investments align to business objectives and value can be realized and quantified. The role is that of an enabler, as a governance overseer, but never to the point of a bottleneck. In some organizations, the COE team is responsible for managing all or part of an infrastructure and the shared data ingestion process, ratifying vendor tools and frameworks for internal consumption. They are either funded directly or get compensated by a chargeback model from the individual lines of business that they service.

The foundational blocks include the following aspects:

- **Cloud strategy**: Which cloud to use, whether a multi-cloud to be considered, and what workloads belong where, depending on special regulatory and compliance guidelines such as FedRAMP

- **Architecture blueprints**: For common data patterns so that reusable assets are created once, hardened, and used multiple times

- **Security and governance**: The deployment model and the mapping of entitlements to principals so that privileges are never misused or misinterpreted

The next set of concerns focus on data ingestion and network connectivity:

- What tools, platforms, and licenses are approved in the areas of ingestion, ETL, streaming, migrations, warehousing, data exploration/visualization, and reporting.

- Individual members from LOB request assistance via a prescribed process such as a ServiceNow ticket.

- Providing scalable operations using **Infrastructure As Code (IaC)**. For example, a data user files a ticket to request assistance in debugging a connectivity issue or enabling a new feature or a new environment. What is the set of prerequisites to facilitate the process?

- Best practices for logging, monitoring, cost, and performance tuning.

- Training and enabling sessions for all data personas. All documentation and training material should be well documented in internal-facing wikis, SharePoint, or similar.

- Building central dashboards where downstream teams can view their usage and billing.

The final set of concerns focus on specialized AI/ML and data science activities:

- Guidance around combining business domain knowledge to power ML activities that are interpretable and explainable. Is business value realized?

- Guidelines around data asset sharing (features data, model, and insights).

- Model acceptance criteria that not only the line of business but also the entire enterprise is accountable for. For example, an insurance model approving loan applications has to be fair and balanced; otherwise, the credibility of the entire organization is at stake.

Usually, a COE is a cross-functional group of members who serve a wide range of users with varying skill sets, from novice data citizens to more advanced players. Their responsibilities fall into the following main buckets:

- Governance and setup

- Deploying infrastructure:

    Different organizations have different policies around the types of environment to provide. Typical ones include the following:

  - **Sandbox**: Users can bring in their own datasets and experiment in a shared environment. Usually, non-sensitive data is allowed.

  - **Developer**: This is typically for a data team and can be shared. Access to data is well guarded.

  - **Staging**: This mimics production. You may have read-only access to production data and access to other secure non-production data.

  - **Production**: Usually, this is off limits to all data teams. All jobs run as service principals and are deployed through CI/CD pipelines.

- Security, monitoring, and alerts/actions:

  - Data and intellectual property loss are the main concerns that are addressed so that people with the right privileges have access to sensitive data through logging, monitoring, and auditing.

  - DevOps and application life cycle management to ensure a high level of automation.

- Approving data projects:

  - Typically, a data team comes with its proposal, which includes what datasets it plans to use, the volume of data, its ETL and consumption strategies, and the use cases it intends to solve. There will be an approximate cost estimate in terms of time, resources and money. There is a score assigned to the project that indicates the priority, COE resources are then assigned to help onboard the team, and a chargeback model ensures that individual teams manage their own operational budgets.

- Nurture and growth:

  - Enable the data community with training and how-tos to facilitate adoption.

  - Provide support to unblock users around tooling and infrastructure.

  - Create reusable assets to facilitate the adoption journey.

- Attain executive sponsorship:

  - This ensures that the vision and mission are clearly articulated with well-defined success criteria. Process efficiencies and accountability are put in place early on.

The establishment of a COE is no guarantee that all data and ML initiatives will be successful, but it is a foundational piece in any enterprise's data journey to ensure a sound governance body for better data management and decision making.

## Summary

The previous chapters focused on the role of data engineers, data scientists, business analysts, and DevOps/MLOps personas. This chapter focused on the admin persona who plays a pivotal role in an organization's data journey by enabling the infrastructure, onboarding users, and providing data governance and security constructs so that self-service can be fully and safely democratized. We looked into various tasks, such as COE duties and responsibilities and data migration efforts, among others, which require admins to do a lot of the heavy lifting. Consolidating data into a common, open format such as Parquet with a transactional protocol such as Delta helps in use cases involving data sharing and migrations. It is important to keep in mind that technology and business users need to plan an enterprise's data initiatives together to make sure that the insights generated are relevant and useful for the enterprise.

# Index

# W

# Z

`Packt.com`

Subscribe to our online digital library for full access to over 7,000 books and videos, as well as industry leading tools to help you plan your personal development and advance your career. For more information, please visit our website.

## Why subscribe?

- Spend less time learning and more time coding with practical eBooks and Videos from over 4,000 industry professionals

- Improve your learning with Skill Plans built especially for you

- Get a free eBook or video every month

- Fully searchable for easy access to vital information

- Copy and paste, print, and bookmark content

Did you know that Packt offers eBook versions of every book published, with PDF and ePub files available? You can upgrade to the eBook version at `packt.com` and as a print book customer, you are entitled to a discount on the eBook copy. Get in touch with us at `customercare@packtpub.com` for more details.

At `www.packt.com`, you can also read a collection of free technical articles, sign up for a range of free newsletters, and receive exclusive discounts and offers on Packt books and eBooks.

# Other Books You May Enjoy

If you enjoyed this book, you may be interested in these other books by Packt:

**Data Engineering with AWS**

Gareth Eagar

ISBN: 9781800560413

- Understand data engineering concepts and emerging technologies
- Ingest streaming data with Amazon Kinesis Data Firehose
- Optimize, denormalize, and join datasets with AWS Glue Studio
- Use Amazon S3 events to trigger a Lambda process to transform a file
- Run complex SQL queries on data lake data using Amazon Athena
- Load data into a Redshift data warehouse and run queries
- Create a visualization of your data using Amazon QuickSight
- Extract sentiment data from a dataset using Amazon Comprehend

**Data Engineering with Apache Spark, Delta Lake, and Lakehouse**

Manoj Kukreja

ISBN: 9781801077743

- Discover the challenges you may face in the data engineering world

- Add ACID transactions to Apache Spark using Delta Lake

- Understand effective design strategies to build enterprise-grade data lakes

- Explore architectural and design patterns for building efficient data ingestion pipelines

- Orchestrate a data pipeline for preprocessing data using Apache Spark and Delta Lake APIs

- Automate deployment and monitoring of data pipelines in production

- Get to grips with securing, monitoring, and managing data pipelines models efficiently

# Packt is searching for authors like you

If you're interested in becoming an author for Packt, please visit `authors.packtpub.com` and apply today. We have worked with thousands of developers and tech professionals, just like you, to help them share their insight with the global tech community. You can make a general application, apply for a specific hot topic that we are recruiting an author for, or submit your own idea.

# Share Your Thoughts

Now you've finished *Simplifying Data Engineering and Analytics with Delta*, we'd love to hear your thoughts! Scan the QR code below to go straight to the Amazon review page for this book and share your feedback or leave a review on the site that you purchased it from.

https://packt.link/r/1-801-81486-4

Your review is important to us and the tech community and will help us make sure we're delivering excellent quality content.